Clonal Forestry II

Conservation and Application

Edited by
M.R. Ahuja and W.J. Libby

With 34 Figures

Springer-Verlag
Berlin Heidelberg New York
London Paris Tokyo
Hong Kong Barcelona
Budapest

Dr. Mulkh-Raj Ahuja
Institute of Forest Genetics
Federal Research Center for Forestry and Forest Products
Sieker Landstrasse 2
W-2070 Grosshansdorf, Fed. Rep. of Germany

Dr. William J. Libby
Department of Forestry and Resource Management
Mulford Hall
University of California
Berkeley, CA 94720, USA

The chapters referring to "Genetics and Biotechnology" are published in Volume I of *Clonal Forestry*, 1993.

ISBN 3-540-55714-8 Springer-Verlag Berlin Heidelberg New York
ISBN 0-387-55714-8 Springer-Verlag New York Berlin Heidelberg

Library of Congress Cataloging-in-Publication Data. Clonal forestry/M.R. Ahuja, W.J. Libby (eds.) p. cm. Includes bibliographical references and indexes. Contents: 1. Genetics and biotechnology – 2. Conservation and application. ISBN 3-540-52501-7 (v. 1): – ISBN 3-540-55714-8 (v. 2) – ISBN 0-387-52501-7 (U.S.: v.1). – ISBN 0-387-55714-8 (U.S.: v. 2) 1. Clonal forestry. I. Ahuja, M.R., 1933– . II. Libby, W.J. SD403.4.C56 1993 634.9′56-dc20 92-25603

Printed in Germany

Typesetting: Macmillan India Ltd., Bangalore 25

31/3145/SPS – 543210 – Printed on acid-free paper

Preface

In the past few decades, much progress has been made in creating the knowledge needed to perform clonal forestry, both in its theory, and in practice. This is reflected in the number and quality of recent scientific and professional meetings that have focused on cloning of forest trees and on clonal forestry throughout the world. In 1988, we decided that a book on clonal forestry would be timely, and we planned it to provide a broad coverage of clonal forestry. However, it later became apparent that such coverage was too much for a single volume, and we have therefore organized a two-volume treatment. This second volume surveys applications of clonal forestry for various purposes in various settings around the world, and the first volume includes much of the science behind such application.

For both volumes, we set out to recruit authors from among the younger scientists in the field, from many countries and regions, to foster a diversity of thought and experience with respect to clonal forestry. When we couldn't resist recruiting some of the longer-established authors, we asked them to write on recent advances or on topics different than those appearing in their previous works.

In our view, cloning forest trees and clonal forestry are not identical. In this respect, we consider the ability to clone forest trees successfully to be a prerequisite to clonal forestry, and much of the recent science and understanding leading to such abilities is contained in Volume 1. Volume 2 provides examples of the use of clones in forestry, and develops the concept of "full clonal forestry".

Briefly, "full clonal forestry" implies that substantial knowledge about the clones being deployed is available and used; that management practices are modified to be clone-specific with respect to such things as deployment, spacing, and harvesting; that genetic diversity within clonal plantations is closely controlled and maintained; and that breeding strategies specifically serving clonal forestry programs may be developed. In short, such developments as the recent advances in genetics, the availability of pedigreed families and clones, and modern methods of data compilation, storage, and retrieval in computers, have combined to evolve new possibilities for clonal forestry, and to begin to create clonal forestry as a discipline. This evolution of clonal forestry theory and practice will no doubt continue. We hope that this Volume will contribute to it, and particularly to responsible planning, thinking, organization, execution, and deployment activities within this new discipline of clonal forestry.

Following a short introductory chapter, Chapter 2 of Volume 2 provides a brief history and an extensive overview of clonal production forestry, and shorter coverage of the place of clonal forestry in urban forestry, agroforestry, and Christmas-tree production. Chapter 3 focuses on both the need for genetic conservation within clonal forestry, and on ways in which cloning can and does contribute to the conservation of forest tree germplasm. We considered this conservation ethic to be so important that it is given equal billing with application in the subtitle of Volume 2.

Chapters 4 to 11 provide case histories of the use of clones in forestry, leading in some of the cases to the practice of clonal forestry in a variety of settings and with a variety of species. Volume 2 closes with a chapter on the organization and implementation of a clonal forestry program, and a chapter reviewing policies and laws affecting clonal forestry.

Each volume is indexed separately, and thus the reader pursuing a particular topic is well advised to check the Indexes of both volumes.

Both volumes have been prepared for a wide range of possible readers, including students, researchers, and forest managers interested in the development of clonal forestry.

Berkeley, CA, USA — W.J. Libby
Grosshansdorf, Fed. Rep. of Germany — M.R. Ahuja
February 1992

Contents

List of Contributors

Addresses are given at the beginning of the respective contribution.

1 Clonal Forestry

W.J. Libby[1] and M.R. Ahuja[2]

1 The Clones Are Coming! The Clones Are Coming!

Techniques of cloning forest trees and, for that matter, of clonal forestry are not new. Cloned trees have been used for a very long time. Kleinschmit et al. (Chap. 2, Vol. 2) provide histories of the use of clones for various forestry purposes, and found evidence for serial grafting of *Ficus* clones for about 3000 years and for air-layering of fruit trees in a 76 A.D. writing. A 12th-century Islamic treatise by ibn-al-Awwam of Seville classified plants as those that grow from cuttings and those that grow from seeds, including in this list more than 50 trees (Hitti 1970). However, with recent advances in genetics and in breeding theory and technique (Namkoong and Kang 1990), with the availability of pedigreed families and clones, with advances in both the theory and practice of vegetative propagation (see *Clonal Forestry*, Vol. 1, Ahuja and Libby 1993), and with modern methods of information storage and retrieval, clonal forestry is evolving as a discipline, particularly during the past decade.

Clonal forestry and some of its specialized topics have been discussed in several major international meetings (Hanover and Keathley 1988; Ahuja 1988, 1991; Table 1, Chap. 2, Vol. 2), as well as in regional meetings (see, for example, Barkley 1985; Miller 1991). There is much interest in clonal forestry (some apprehensive, some enthusiastic) in organizations involved in the genetics, breeding, and improvement of forest trees in many parts of the world.

We, and most of those who have joined us in writing these two volumes, think that the advantages of clonal forestry are attractive and its problems manageable. Furthermore, its technical problems are yielding to research and understanding (see Vol. 1), and various strategies are being tested and refined in pioneer programs with a variety of species and in a variety of settings. Volume 2 reviews clonal forestry's advantages and problems, and presents clonal forestry as it has been long practised with some species, as it has been recently implemented with others, and as it is being planned for still others. Given that a few as-yet-unproved assumptions turn out to be sufficiently correct, among them that civilization survives, then it seems likely that on our more productive sites, particularly those closest to mills, but in other settings as well, clonal

[1] Department of Forestry and Resource Management, Mulford Hall, University of California, Berkeley, CA 94720, USA

[2] Institute of Forest Genetics, Federal Research Center for Forestry and Forest Products, Sieker Landstrasse 2, W-2070 Grosshansdorf, FRG

Clonal Forestry II, Conservation and Application
Ed. by M.R. Ahuja and W.J. Libby

forestry will increasingly be practiced in the near future, and it will be the common form of intensive forestry in the more distant future.

2 Clonal Forestry and Related Definitions

Clonal forestry as a general system was proposed in 1906 by G. Andersson in Sweden, followed by similar proposals in Germany by H. Fabricius in 1922 and in Denmark by A. Oppermann in 1923. All three proposed using grafting on a wide scale, following horticultural practice at the time. But the care and attention focused on trees by horticulturists, particularly with respect to propagule cost and knowledge about specific varieties or clones, were generally considered inappropriate for forestry (Larsen 1956). This may have now changed.

"Cloning forest trees" and "clonal forestry" sound very much alike, but there are some important differences. The primary focus of Volume 2 is on the applications of clonal forestry. In spite of its long history, this is essentially a new way to work in forestry, providing many opportunities not available with classical seedling-based methods. We take the position that "clonal forestry" implies much more than just using cloned trees in forestry. Being able to clone trees is, of course, prerequisite to doing clonal forestry, but using cloned trees in forestry practice is not, in our opinion, sufficient to qualify as practicing clonal forestry.

2.1 Some Examples of Cloning Forest Trees

Cloning forest trees has been widely practiced during the past four decades or longer, for a variety of purposes. These include:

1. To clone parent trees and their contained packages of genes into seed orchards or breeding orchards, for multiplication and redistribution of their genes into the next generation (see, for example, Chap. 11, Vol. 2).
2. For non-genetic research purposes, to attain greater uniformity and repeatability (Chap. 6, Vol. 1).
3. For genetic research purposes, to be able to investigate whole genotypes replicated in space and time (Chap. 2, Vol. 1).
4. For "tree shows", which are meant to effectively demonstrate genetic differences between trees of different genotypes (Larsen 1956).
5. For genetic conservation purposes (Chap. 3, Vol. 2).
6. As a preferred method of propagation. For instance, some species of willow and poplar are more cheaply and effectively propagated from cuttings than from seeds (Chap. 5, Vol. 2).
7. More recently, to clonally "bulk up" (vegetatively multiply) inadequate numbers of seeds or seedlings, from appropriate provenance collections, seed orchards, or controlled crosses, to numbers sufficient for production plantings (see Sect. 2.4 below).
8. The use of vegetatively multiplied material from controlled crosses can avoid

problems associated with open-pollinated seed-orchards such as selfing, pollen contamination, imbalanced matings, and maladapted matings (Libby 1983).

9. By the use of vegetatively multiplied material from ortets that have matured to some degree, one can capture some favorable developmental changes that are known to occur in the clones of some species with maturation (Libby 1990; Chap. 3, Vol. 1; Chap. 8, Vol. 2).

However, any or all of the above, unless some additional dimensions are added, do not constitute clonal forestry.

2.2 Characteristics of Clonal Forestry

So, what is clonal forestry? The following are some characteristics that, singly or in combination, distinguish clonal forestry from merely using cloned forest trees:

1. First, something must be known about the clones. This implies clonal testing (Chap. 9, Vol. 1). Associated with this is that the clones are selected based on their performance as clones, and not merely by being members of proven or promising families, populations, etc. As more test-based knowledge about the deployed clones is obtained and used, more advantages of clonal forestry accrue. For example, with short-term testing of candidate clones on one site, one learns the relative performance among clones for some early-expressed traits. For long-term testing of either candidate or production clones on many sites, one also learns about near-rotation-age traits, whether there are differences among clones in adaptation to different sites, whether some clones are broadly adapted while others are more site-specific, and other things that make their deployment, silviculture, and uses more efficient and effective.

2. Another distinguishing characteristic of clonal forestry is that well-known, reliable clones are repeatedly used, often over many vegetative generations or harvest cycles. The Lombardy poplar is such a clone that has long been used effectively for various urban and amenity purposes.

3. Under clonal forestry, management practices do (or at least could) adapt to particular clones, from details of propagation and nursery management, through deployment assignments and silvicultural treatments and schedules, to harvest sequence and end uses.

4. A common feature of clonal forestry is that initial spacing between trees in production plantations tends to be greater than that employed with sexually produced seedlings or with clonal planting stock "bulked up" from mixtures of heterogeneous seedlings. This flows from the greater reliability and uniformity of identified tested clones, and from the confidence that is based on management experience with well-known clones. It is also commonly due to selection for favorable tree form during clonal testing. The money saved by planting fewer trees and on subsequent thinning costs can be shifted to other management options, such as pruning, fertilization, irrigation, and/or higher profit margins.

5. Useful predictability may be obtained both for single well-known clones and for combinations of clones (see for example the urban forestry section of Chap. 2, Vol. 2).

6. Diversity can be more carefully programmed into a clonal plantation consisting of a relatively few well-known pedigreed clones (Libby 1982; Chap. 4, Vol. 1), compared to "bulked up" or sexually propagated plantation material. This dimension of clonal forestry is just beginning to be practised, for example with eucalypts in South America (Chap. 7, Vol. 2). It is a topic receiving considerable attention by both policy-makers and legislators in Europe and Canada (Chap. 13, Vol. 2), and by researchers in many places.

7. Additional genetic gain may be captured. It is commonly recognized that clonal selection can capture nonadditive genetic variation, and that favorable nonadditive combinations are mostly lost with sexual propagation. It is less commonly recognized that, even with full-sib families, half of the available additive genetic variation is segregating within those families. Individuals in such families that are outstanding due to favorable combinations of additive genetic variation can be clonally propagated (Libby 1983).

8. Some novel options become available (Libby 1987), such as deploying trees that devote little energy to sexual reproduction (hard to do if they are seed-orchard offspring), or growing the same amount of wood on shorter fat trees, rather than on taller skinny trees (hard to do if the planting mixture contains both).

9. One may select for correlation breakers (such as clones with both above-average diameter and above-average specific gravity, of possible importance in radiata pine clonal forestry).

10. One may utilize maturation states other than juvenile at the time of planting (there are some important advantages in this for radiata pine, see Chap. 8, Vol. 2).

11. One can also keep specialty genotypes around for occasional use. If seedlings are used for production planting, it is likely to be counterproductive if the specialty parents are in a general-purpose open-pollinated seed-orchard. (A two-way pollen contamination would probably result, since pollen of the specialty clones is likely to be inappropriate for the general purposes of the seed-orchard, and pollen from the other seed-orchard parents is similarly inappropriate for the specialty purposes). Even if trees for occasional use were kept in a control-pollinated breeding-orchard, the lag time between order and delivery might be several years. Compare these two seedling options to hedges of the specialty clones trimmed to waste in most years but used as cutting donors in some, or to tissue-culture storage of occasionally used clones.

12. A developing opportunity is to provide material from very good genotypes for insertion of single genes, to add some desired feature or to correct some defect in otherwise excellent performance (Chaps. 10–15, Vol. 1). When this is done with already well-known clones, it makes sense to clonally deploy the modified genotype rather than sexually propagating the inserted gene into a segregating breeding-line or population and then deploying the heterogeneous progeny.

13. Finally, within-clone uniformity in many important traits (size, form, growth requirements, and wood) is a major advantage of clonal forestry (see Chap. 7, Vol. 2). On this point, it is often recognized that clonal forestry will lead to genetically uniform traits, but then assumed that this will also lead to undesirably low levels of genetic diversity in a plantation or region. Poor

management decisions can result in such a mistake, but it is not a built-in disadvantage of clonal forestry. On the contrary, genetic diversity can be effectively maintained or even enhanced in clonal forests by deploying a sufficient number of unrelated elite genotypes (Libby 1982, 1990). The number sufficient to create a clonal forest that is as safe as a natural seedling forest is in the range 7–30 unrelated clones. More than 30 clones may actually increase the risk of some kinds of epidemics, and also dilute or lose several of the advantages of clonal forestry.

2.3 Family Forestry

The concept and early practice of "family forestry" is largely a development of the past few years, with a bit of earlier pioneering in the southeastern U.S.A. In this, tested orchard-open-pollinated, polycross or full-sib families are collected and raised separately, and then are deployed as single families or as prescribed family mixtures to particular compartments or sites. Family forestry incorporates wholly or in part many of the advantages of clonal forestry, and in some cases it may provide an attractive alternative to clonal forestry (Carson 1986; Libby 1990).

2.4 Vegetative Multiplication

Vegetative multiplication (VM) is essentially synonymous with "bulking up", as used to describe cloning of seedlings in short supply to increase the numbers of propagules for deployment to plantations. VM allows many-fold expansions from seedlings or seeds to deployed stecklings, plantlings or emblings. It is usually done at a juvenile state, but advantages of (for example) the adolescent state can be captured by delaying VM until the ortets or donor ramets are at that maturation state.

2.5 Zygotic Forestry

Whether the deployed propagules are genetically unique seedlings or vegetatively-multiplied propagules, the important criterion of zygotic forestry is that the individual genotypes are not characterized or well known. Performance knowledge focuses on the family, line, breed, or population, but management cannot take advantage of differences among clones, or of extensive knowledge about clones.

3 The Interdependence of Cloning and Breeding

Clearly, clonal forestry and breeding are more than just compatible (Zobel and Talbert 1984). An appropriate breeding program should be maintained in support of a clonal forestry program, with outstanding candidates from each new generation tested against the current "first team" of deployed clones. The recent-generation candidate clones will replace the long-term production clones as test results show some of them to be sufficiently improved. One should note that "sufficiently improved" is probably not just a little bit better. The candidate clones must be substantially better in order to compensate for the very real value of management experience with the "first team" production clones.

However, there is an important contest of strategies between the practice of clonal forestry and that of relying on the products of continued breeding (see Chaps. 2 and 5, Vol. 1). Very aggressive breeding, using early testing and rapid generation turnover, may prove to be so successful that it routinely produces average offspring better than the best clones of a few generations back. If so, these new breeds, or the better clones among them, will replace the older clones before test information and management experience accumulate sufficiently to justify full clonal forestry. Thus, whether the full benefits of clonal forestry are likely to be attained or not will depend on how well aggressive breeding works. Experience with grapes, fruit trees, poplars, etc indicates that clones currently in use, simultaneously combining many desired characteristics, are really much better than the average of later-bred generations. Once accepted in practice, these outstanding "first team" clones are tougher to displace than one might expect. In many, perhaps most, forest-tree species, we may find the same to be true of the elite super clones that our testing and practice will find and accept in the coming decades. This ability of excellent well-known clones to remain in widespread use is a necessary criterion for clonal forestry to replace the more traditional zygotic forestry, which has been based on the average performance of highly heterogeneous planting stock.

4 Conservation

Chapter 3 of Volume 2 presents thorough overviews of why genetic conservation of forest trees is a matter of high priority, and of ways to accomplish such conservation. It gives particular attention to how cloning techniques can and do contribute to genetic conservation programs. Genetic conservation should be a significant component of any large forest-management program. It has special ethical and practical relevance for those engaged in the development and/or implementation of clonal forestry, as the widespread use of clones poses both a perceived and a real threat with respect to the amount and structure of genetic diversity in our future forests. We hope that those engaged in these aspects of clonal forestry will be informed as to the importance of genetic conservation, and that some of them will be among the leaders in addressing and responding to these issues. If, instead, the pioneers of clonal forestry are viewed by conservationists as a problem, or even as the enemy, then support for the development and implementation of clonal forestry may erode, and important (and we hope unnecessary) opposition may develop.

5 The Beginnings of Clonal Forestry

Clonal forestry has been practised the longest with sugi (Chap. 4, Vol. 2), and over the greatest areas with poplars and willows (Chap. 5, Vol. 2). We have much to learn from the successes and from the mistakes of these programs. For example, with respect to poplars and other dioecious species, it is sometimes noted that more male clones are selected (usually in tests employing monoclonal blocks of less than a hectare) for productivity per unit-area than are female clones. It occurs to us that the females may be allocating a greater proportion of available photosynthate to fruit and seeds than are the males to anthers and pollen (see review in Libby 1987). If so, and if failure to pollinate results in abortion of most of the female structures early in development, then some greater proportion of the available photosynthate in females might be reallocated to stem-wood. Deployment of female clones in large monoclonal blocks, or in smaller adjacent monoclonal blocks, or in intimate mixture with each other over large areas, might sufficiently reduce pollination to have such an effect.

6 Present and Future Practice of Clonal Forestry

In British Columbia, cloning for production deployment is being attempted in seven species and practiced to varying degrees in four of them (Chap. 11, Vol. 2). However, the programs are young, and it is early days for clonal forestry. Nevertheless, the B.C. program may claim one important landmark, namely, its program with yellow-cedar is the first with a conifer in which more than half the annual planting on a production basis used stecklings rather than seedlings. By 1988, over a half million yellow-cedar stecklings were being produced and planted annually by (mostly) private growers.

Perhaps the greatest recent increase in activity utilizing full clonal forestry is in tropical and subtropical regions with the genus *Eucalyptus*. We originally requested a general overview of eucalypt clonal forestry, but our author had to withdraw at the last minute for health reasons. Bruce Zobel stepped in and responded immediately with a chapter focused on clonal forestry with eucalypts in South America. We are grateful for this, as the eucalypt clonal forestry story is one of the best examples of the development of this new kind of forestry. As might be expected with Dr. Zobel, Chapter 7 contains some insights about wood and wood quality that have generality well beyond the clonal forestry of eucalypts.

Once forest-tree DNA has been genetically engineered and incorporated in plantable propagules, these have to be tested, and the successful ones then deployed on a very large scale to recover the research and development costs. Before such commercial application of genetic engineering is developed, be it for clonal forestry or for zygotic forestry, it seems likely that rules and laws governing the release and international trade of this material will be developed and enforced (Chap. 13, Vol. 2).

Many of the topics in 2.2, above, are illustrated repeatedly in Chapters. 2–13, of this Volume. These chapters have been assembled to provide insights into the history of clonal forestry; into several ways of practicing it in different settings, with different species, and for both general and specific purposes; and into ways in which it has been, is being, and may be organized and regulated.

In summary, we consider clonal forestry to be a complicated discipline that has a thin history, an exciting present, and a bright future. To the degree that it will allow more effective management of some of our forests, it will not only provide *Homo sapiens* with a renewable supply of wood and other forest commodities and benefits, but it will also allow greater areas of natural forest to be reserved as a result of these greater efficiencies, thus serving and perhaps saving thousands or even millions of Earth's other species as well.

References

Ahuja MR (1988) Somatic cell genetics of woody plants. Martinus Nijhoff. 422 pp

Ahuja MR (Ed) (1991) Woody plant biotechnology. Plenum Press, New York. 373 pp

Ahuja MR, Libby WJ (1993) Clonal forestry. Vol. 1. Genetics and biotechnology. Springer, Heidelberg

Barkley BA (1985) Proceedings fast growing hardwoods workshop. Brockville, Ontario, Canada

Carson MJ (1986) Advantages of clonal forestry for *Pinus radiata* – real or imagined? N Z Jour For Sci 16:403–415

Hanover J, Keathley D (Eds.) (1988) Genetic manipulation of woody plants. Plenum Press. 519 pp

Hitti PK (1970) The arabs. A short history. Revised paperback edition. Gateway Editions. Chicago, 274 pp

Larsen CS (1956) Genetics in silviculture. Oliver & Boyd, London. 224 pp

Libby WJ (1982) What is a safe number of clones per plantation? pp. 342–360. In: Heybroek HM, Stephen BR, Von Weissenberg K (Eds.) Resistance to diseases and pests in forest trees. Purdoc. Wageningen. 503 pp

Libby WJ (1983) The clonal option. Norsk. Institutt for Skogsforskning. 1432 As-NLH, Norway. 32 pp

Libby WJ (1987) Do we really want taller trees? Adaptation and allocation as tree-improvement strategies. The H. R. Macmillan Lecture in Forestry. Univ of British Columbia, Vancouver. 15 pp

Libby WJ (1990) Advantages of clonal forestry revisited. Sec. 5.13, 13 pp. In: Proc Joint Meeting of WFGA and IUFRO Working Parties S2.02–05, – 06, – 12 and – 14. Olympia Wa. 20–24 August 1990, Weyerhaeuser Co., Tacoma, Wa

Miller JT (Ed) (1991) Proceedings FRI/NZFP Ltd Clonal Forestry Workshop, 1–2 May 1989, Rotorua, New Zealand. Ministry of Forestry (N Z) FRI Bull. 160. 200 pp

Namkoong G, Kang H (1990). Quantitative genetics of forest trees. Plant Breeding Reviews 8:139–188. Timber Press, Portland Or

Zobel BJ, Talbert JT (1984) Applied forest tree improvement. J. Wiley & Sons, NY. 505 pp

2 Past, Present, and Anticipated Applications of Clonal Forestry

J. Kleinschmit[1], D.K. Khurana[2], H.D. Gerhold[3], and W.J. Libby[4]

1 Introduction

Trees are among the few natural resources that can be renewed, and the need for trees and the wood that they provide is continuing to increase with increases in world population and the striving of peoples to maintain or increase their standards of living. The supply of wood is threatened by overcutting in many regions, and in some areas the near-complete destruction of natural forests threatens other values. More intensive and aggressive forest management can improve the supply of wood and other forest products and values on existing forest land, and on marginal or abandoned farm land that can be afforested. Vegetative propagation and the appropriate deployment of clones can contribute considerably to increases in the productivity and effectiveness of such managed forests, of agroforestry ecosystems, of Christmas-tree farms, and of trees for landscape, urban and amenity uses. Interestingly, tree-breeding programs can become much more flexible when serving clonal forestry than when restricted to the production and deployment of seedlings (Libby and Rauter 1984).

As elsewhere in this Volume (Libby and Ahuja, Chap. 1, for example), we use the concept "clonal forestry" here in the sense that clones are deployed in ways that lead to various advantages for forestry purposes. Many of these advantages are less available or unavailable if seedlings or untested clones are deployed. Under full clonal forestry, each clone has been well tested, and thus deployment of such clones is based on reliable knowledge and expectation of clonal performance.

2 Background

Most forest managers have traditionally preferred to plant or replant their forests using seedlings resulting from sexual reproduction. Perhaps from a very

[1] Lower Saxony Forestry Research Institute, Department of Forest Tree Breeding, W-3513 Escherode, FRG
[2] Department of Tree Breeding, University of Horticulture & Forestry, Nauni, Solan (Himachal Pradesh) PIN 173 230 India
[3] School of Forest Resources, Penn State University, University Park, PA 16802, USA
[4] Department of Forestry and Resource Management, University of California, Berkeley, CA 94720, USA

Clonal Forestry II, Conservation and Application
Ed. by M.R. Ahuja and W.J. Libby

human point of view, they have regarded vegetative propagation of both people and trees with a certain suspiciousness. Yet clonal propagation occurs in nature, producing identical twins, triplets, quadruplets, and quintuplets in humans, and even more extensive natural cloning occurs in some species of trees (see Rydelius and Libby, Chap. 9, Vol. 2). For example, aspen and wild cherry propagate by root suckers; spruce, pine and Port-Orford-cedar sometimes layer, i.e., branches buried in litter or moss form roots and become independent plants; redwoods and some eucalypts sprout from the root collars of cut or burned trees, forming circular clonal groves; and detached branches of poplar and willow can root and grow at some distance from the ortet tree.

In recent years, it has become clear that at least some cells in most plant species are totipotent, i.e., they contain all of the information needed to regenerate a complete plant. This property allows the cloning of plants. There are several ways to do this, discussed briefly in Section 3.1, below, and more extensively in Chapter 7 (Chen and Ahuja) and (Talbert et al.), Volume 1. The number of species for which effective vegetative propagation is now possible is increasing each year, as greater understanding of plant development accumulates (see Greenwood and Hutchison, Chap. 3, this Vol. 1) and as various techniques are attempted, modified, and optimized.

Some of the advantages of using vegetative propagules for various forestry purposes are presented in Ahuja and Libby, (Chap. 1, this Vol.) and details of many of them appear in Chapters 4 through 12. Three advantages that frequently convince management to use clonal material are: (1) Where or when a shortage of appropriate seeds exists, a limited supply of such seeds can be clonally expanded to meet planting needs. (2) Outstanding phenotypes that combine several desirable properties are occasionally noted among the many trees in plantations or in the forest. Managers have become aware that, by cloning these unusual outstanding trees, it is possible to produce large numbers

Table 1. Meetings and Symposia held 1973–1983

Meetings	References
1: 1973. IUFRO Meeting on Vegetative Propagation. Rotorua, New Zealand.	IUFRO 1974
2: 1975. Symposium on Juvenility in Woody Perennials. College Park, USA/Berlin, Germany.	Symposium 1976
3: 1977. Meeting on Vegetative Propagation of Forest Trees – Physiology and Practice. Uppsala, Sweden.	Swedish 1977
4: 1981. Symposium on Clonal Forestry. Uppsala, Sweden.	Swedish 1981
5: 1981. Meeting on In vitro Propagation of Forest Tree Species. Fontainbleau, France.	AFOCEL 1982
6: 1982. IUFRO Meeting About Breeding Strategies Including Multiclonal Varieties. Sensenstein, F.R. Germany.	IUFRO 1982
7: 1982. Improving Vegetatively Propagated Crops. Abbott AJ, Atkin RK (eds).	Abbott and Atkin 1987
8: 1983. Clonal Forestry: Its Impact on Tree Improvement and Future Forests. Zsuffa L, Rauter RM, Yeatman CW (eds.)	Zsuffa et al. 1985

of plants that are genetically identical and phenotypically similar to them. (3) Managers are recognizing that clones offer substantial efficiencies in planning and executing various forestry activities. This is especially based on the expectation that clones are more predictable and more uniform (within clones) in their growth and development than are seedlings (see Libby and Rauter 1984; Libby 1990 for details of these and other management efficiencies associated with clones).

These advantages, plus others that are more specialized or that usually occur to managers later, led to abundant research into the techniques and applications of vegetative propagation. It was during the period of roughly 1972–1983 that the possibilities of widespread application of clonal forestry with many species were explored and developed. The *Proceedings* of eight meetings and symposia that were held during the period 1973–1983 are listed in Table 1.

2.1 The Development of Clonal Material

Clonal material has long been selected and distributed in horticulture. In forestry as in horticulture, selection was commonly done by experienced "green-fingered" practitioners, for example in sugi (*Cryptomeria japonica*) Ohba, Chap. 4, Vol. 2) and poplars (Zsuffa et al., Chap. 5, Vol. 2). Outstanding clones of these early poplar selections are still traded worldwide. Systematic breeding and selection with an integrated clonal testing program is more recent in poplars and in a few other tree species. Some selection procedures are discussed in Talbert et al. (Chap. 11, Vol. 1).

Classically, forest-tree species were propagated either vegetatively or generatively (zygotically). Even if both options were used in parallel, as with sugi, there was little integration of the two systems. This resulted in considerable differences in performance and productivity between vegetative and generative material. For example, zygotic poplar propagules have no market in production forestry at all. Ernie Schreiner, one of the early poplar breeders, was a powerful advocate for breeding that was integrated with clonal propagation and deployment (Schreiner 1938, 1970), and he was perhaps the first to recommend the use of clonal mixtures.

In some of the more recent programs, both options are closely integrated (see Bentzer, Chap. 6, Vol. 2). Deployment of vegetative propagules is used for immediate realization of gain, but clonal tests often double as breeding orchards, and genes of the best clones are thus forwarded in the generative cycle. This option can broaden the genetic base of the breeding population considerably. For example, in the spruce-breeding program of the Lower Saxony Forest Research Institute, 60 000 clones have been selected and included in clonal tests. The origins of these clones are diverse, from many different populations. If the top 10% of these clones are included in the next breeding cycle, after 15–20 years' testing time, and if they are combined by single-pair matings, 3000 full-sib families will be produced. Sublining is easily possible if such controlled crosses are made. A similar principle holds for species-hybrid programs. The (often expensive) species hybrids can be vegetatively multiplied and deployed to production plantations, as is being done in poplars, eucalypts, and spruces. Clonal tests of the species hybrids can be used to select outstanding clones and then to produce F_2 or more complex three- and four-species hybrids. In contrast to agriculture, both parents of the F_1 hybrids are typically highly heterozygous, and selection among sibs for growth and adaptation is effective in the F_1 generation.

Under practical forestry conditions, this means that vegetative and zygotic material can both be effectively used. The two kinds of propagule may even be grown in specific programmed mixtures, for example by interplanting seedlings or by allowing natural regeneration between outstanding clones planted at wide (harvest) spacing. This may diminish some risk, such as uncertainties about maturation of the clones, and it will diminish the cost of using 100% clonal material at close spacing. This needs new thinking and perhaps new concepts in practical forestry.

2.2 Recognized Advantages of Clonal Forestry

We divide this section into two parts, one on present advantages of using clones for various applied forestry purposes, the other on present advantages of installing clonal forestry practices (see Libby and Ahuja, Chap. 1, Vol. 2, for a discussion of the differences).

2.2.1 Some Advantageous Uses of Clones in Applied Forestry

A continuous or reliable supply of plantation stock is a major problem for practical forestry in species such as oaks, beeches and dipterocarps, which have large seeds that cannot be effectively stored. In other species, and in particular in some tropical species, the problem is more one of reliable seed acquisition from scattered and/or erratically flowering trees in nature. In some of these species, these problems have prevented the development of plantation forestry. For example, successful vegetative propagation has helped to overcome these difficulties in *Triplochiton scleroxylon* (Leakey 1987), *Nauclea diderrichii*, and *Khaya ivorensis* (Mathias 1988), and at the same time such clonal propagation allows some selection and thus improvement of these difficult species.

Controlled crossing may produce valuable seeds (i.e., genotypes), but because it is time-consuming and costly, these seeds may be produced in limited quantities. As an example, testing may have established the outstanding performance of a few full-sib families, but their practical use is limited by the quantity of flowers on the breeding orchard trees, or by a shortage of labor or funds to produce adequate numbers of seeds by controlled crosses. But modest numbers of such high-value seeds or seedlings can in some species be easily and rapidly "bulked up" for practical deployment by various methods of vegetative propagation. This vegetative multiplication of tested families provides an option intermediate between that of control-pollinated (c-p) seed-orchards (to produce large volumes of seeds of specific full-sib and polycross families, see Carson 1986) and full clonal forestry (where each clone is tested and deployment is based on clonal performance). Note that this intermediate stage requires cloning facilities to be built and cloning problems to be solved. It thus can serve as a bridge between zygotic forestry and full clonal forestry. Once the facilities are in place, and experience in cloning accumulates, the step to full clonal forestry will be relatively easy and often compelling.

2.2.2 Some Advantages of Clonal Forestry

Once identified, high-yielding and/or high-quality clones can be quickly deployed in large numbers per clone for practical forestry purposes. This is a significant advantage of clonal forestry, recognized and applied with poplars, eucalypts, spruces, cherry, and pines. Modern methods of vegetative propagation have allowed large-scale deployment in only a few years after the clones are initially selected. This is much faster than is generally possible using the sequence of selection, seed-orchard establishment and seed production that is typical of a classical tree-improvement program. This time difference is largely due to the several years it takes most trees to reach a size and (sometimes) maturity necessary for pollen and seed production at the levels required for production plantations. In addition, by using selected clones, one may avoid the disadvantages associated with sexual recombination in seed-orchards; namely, that pollen contamination, some selfing, and unbalanced matings often occur in o-p seed-orchards, that even c-p seed-orchard full-sib families capture only a small fraction of the non-additive genetic variation that may have contributed to some parents' outstanding performances, and that 50–75% of the additive genetic variance occurs as segregating genetic variation around each family's mean.

No other form of propagation offers the possibilities of using specific adaptations or trait combinations as efficiently as does clonal propagation. Clones that are late (or early) flushing, that have a high ratio of root dry-matter to total dry-matter, or narrow (or broad) crowns, or high (or low) wood density, can be selected and then mass-produced for appropriate deployment. If a broad base of clones is available, sets of specific clones for various specific uses can be propagated without the long-term investments that are necessary for producing a diverse set of sexually propagated lines or varieties. This gives the breeder and the silviculturist great flexibility.

A special case of specific adaptations involves the spatial deployment in production plantations of complementary clones. For example, a systematic mixture of fast- and slower-growing clones can accomplish natural pruning and minimize competition effects on the final crop trees. Such mixtures can predetermine thinning activities (Fig. 1). By avoiding the accidental clumping of random groups of the better and the poorer producers, the entire plantation site can be utilized by genotypes destined to be crop trees, and uneconomic thinnings within clumps of either wholly better or wholly poorer producers can be avoided.

Another option might be an intimate and systematic mixture of different flushing types, with a goal of increasing productivity of the stand per unit area in

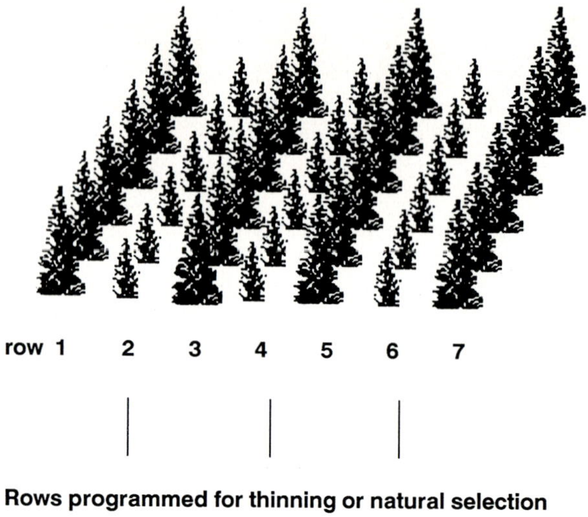

Fig. 1. Example for systematic planting pattern

most or all years. In years without late spring frosts, early-flushing clones use the site better; in years with late spring frosts, the late-flushing clones have an advantage and contribute more to overall productivity.

Desired traits sometimes have additive genetic correlations that are negative. For example, trait pairs such as diameter growth and wood specific gravity or juvenile-height growth and date of bud set sometimes have such negative correlations. These genetic correlations predict the outcome of sexual recombination in nature or in a seed-orchard, but they are rarely perfect (that is, r is rarely -1.0). In most populations, individuals can be found that combine above-average performance for both traits. An effective way to capture these "correlation breaker" combinations is by clonally testing those exceptional individuals that exhibit them, and then clonally deploying the candidate clones that prove to have such combinations.

In natural forests, the forester has little control of its genetic diversity in the course of one or two generations, although diversity may be affected (usually negatively) during longer periods of forest management. The genetic diversity of plantations can be substantially increased or decreased, compared to that found in nature, depending on the species mix and the genetic architectures of the populations used in the plantation. If seedlings are used, the seed donors may have been only a few trees, or they may include a sample of the entire gene pool of the species. Even if a seed-donor stand contains a genetically broad sample, the collected seeds may come from a small part of the available population. For example, in a given year a late frost might kill the strobili of all but the very late flushers, or, even worse, year after year the seed donors might be a few broad-crowned easily climbed trees. Many examples of such situations exist in practical plantation forestry where the seeds are harvested from natural stands or previous plantations. Well-designed seed-orchards allow a particular level of genetic diversity to be programmed into the potential seed production, and orchard management can at least monitor and, perhaps through cultural practices, even modify the effects of such events as late spring frosts. But the best control of plantation diversity is provided by clonal forestry, where sets of pedigreed clones of known genetic relationships and known performances can be prescribed for each plantation. The range of options is bounded by a single genotype on the low end, exemplified by the extensive planting of poplar clone I-214 in the Danube, Drava, and Sava River basins in Yugoslavia (Guzina 1981), and it is unbounded on the high end, with clonal diversity substantially exceeding that of most seedling populations when heterozygous clones of different origins are deployed. When such contrasting clonal diversity is deployed, it can also serve such goals as "biological automation" of thinning, better utilization of specific biological niches, and/or, by understanding the compensatory demands of neighboring clones as discussed above, increasing unit-area productivity.

2.3 Recognized Problems of Clonal Forestry

The practical application of clonal forestry is still in its infancy; as such, it is reasonable to expect that it will still have many problems. The scale of such application is often large, the values high, and the up-front costs substantial. Thus, it is expected that foresters will be cautious in adopting such radical new technology, even if the advantages are tempting. We discuss some recognized problems and concerns associated with clonal forestry, surely not an exhaustive list, under four headings below. Although there are serious economic, technical, and ethical problems with clonal forestry, at present the biological problems seem to be the ones of most concern.

2.3.1 Biological Concerns

Some have considered the most serious and obvious biological disadvantage of clonal forestry, in comparison to zygotic forestry, to be a possible reduction of genetic variation. This "diversity problem", however, is not really a biological problem; rather, cloning technology provides a greater and more serious opportunity for errors in deployment strategy by management. Thus, the solution lies in the education of forest managers, or perhaps in wise regulation (Muhs, Chap. 13, Vol. 2). The consequences of each manager independently deciding to deploy only the very best few clones of those available must be made clear and perhaps avoided through appropriate regulations. Biological research into measures of genetic diversity and into pathosystem epidemiology, for example, will better allow us to understand the importance of diversity and to devise strategies to deploy adequate diversity in clonal plantations. Appropriatc deployment of even a modest number of diverse clones may actually increase the genetic variation within clonal plantations compared to similar plantations of seedlings, and thereby reduce the relative risk of loss in clonal plantations to below that of seedling plantations (Libby 1982).

Even if risk in individual plantations or in local regions is well handled by the deployment of a diverse set of clones, there is still the severe problem of erosion of the overall base of genetic diversity in a species. If plantations make up only a modest fraction of the area occupied by each component species in a region, the continued evolution of naturally reproduced stands in the region should maintain adequate and appropriate reserves of genetic variability. However, this will not be the case if the trees left to reproduce following logging within such naturally reproduced stands are left on some basis other than random. Such stands will respond by becoming landraces that no longer contain the full range of native genetic variation. If most or all native forests are replaced by plantations, as is the case with Norway spruce in southern Sweden and with native mixed hardwood forests in several tropical countries, even a broad base of families or clones being deployed from seed-orchards or in clonal forestry will be a biased and perhaps inadequate sample of the native genetic variation. Increasingly, genetic conservation is being recognized as an important part of most tree-breeding programs; this philosophy is, if anything, even more appropriate and important for clonal forestry programs (Millar, Chap. 3, Vol. 2).

With a few exceptions (poplars, sugi, *Chamaecyparis*, some landscape trees), the performance of clones being deployed in clonal forestry is known to rotation age only for short-rotation uses. For long-rotation production forestry, clones now being deployed have been selected based on their performance for only a fraction of the rotation. It is therefore possible that problems may occur later in the rotation. This is one argument for using intimate clonal mixtures or mixtures of clones and seedlings. Arguments against deployment in monoclonal stands thus include not only the likelihood of lowered productivity due to intense competition among genetically identical neighbors (Harper 1977), but the risk of total loss of some monoclonal blocks at an awkward time in the rotation – namely, at a stage beyond the stage at testing, but when the trees are still too small for effective salvage. This problem of short-term testing of trees destined for large-scale long-term deployment is also a problem with agronomic species in which uniformity is a goal. The difference between most crops and forest trees is the long commitment of each tree and stand between planting and harvest, which does not allow a rapid change in the set of clones or varieties deployed in the region at risk. For example, following the outbreak of corn blight in the United States, most farms had replaced susceptible corn varieties with resistant varieties in two years. But, following the stem-canker and leaf-rust outbreaks on poplar clone I-214 in Yugoslavia, 20 years' worth of monoclonal plantations of I-214 could not be so quickly replaced. Improvements in early testing or greater patience to wait for the results of longer-term testing will help. But even rotation-age testing is not a full solution if it is done on a small scale. Rare but virulent pathogens and pests that may easily miss inclusion in small-scale tests may encounter the new clone or variety when its large-scale deployment is practiced, and then multiply and produce a catastrophic epidemic. Or, a previously mild form of a pest may evolve into a virulent form when enough "clonal substrate" is deployed to favor the evolution of that pest to a form able to attack this newly abundant clone. The best defense with respect to this problem seems to be the wise deployment of sufficient genetic diversity to allow some expected and acceptable losses while reducing the chance of catastrophic loss (Libby 1982; Lindgren, Chap. 4, Vol. 1).

In some taxa, such as the poplars, maturation effects are generally believed to be of little practical importance (Dunlap 1990). In others, such as radiata pine, maturation effects can on balance be advantageous for some practical applications (Gleed, Chap. 8, Vol. 2). But for most taxa, maturation effects provide vexing impediments to the application of cloning in practical forestry (Greenwood and Hutchison, Chap. 3, Vol. 1). The problems are twofold. First, vegetative propagation from juvenile ortets is generally fairly easy, but for some taxa it may be prohibitively difficult from more mature ortets, or from ramets of originally juvenile clones as they mature during testing and other use. Second, even if vegetative propagules can be produced from more mature donors, their subsequent growth is often substantially different from that of juvenile propagules, and the differences are often unfavorable, on balance, or at least the performance of the more mature propagules is sufficiently uncertain so that foresters are slow to commit to the use of such clonal material. Progress in understanding and dealing with these problems is presented in Bonga and von Aderkas, Chap. 12, Vol. 1.

Within-clone genetic variation may occur due to somatic mutations, particularly in large clones or in clones that have been repropagated for many cycles. Somaclonal variation may occur with even higher frequency among propagules of cell- or tissue-culture origins (Chen and Ahuja, Chap. 7, Vol. 1). So far, this seems more a concern than a clearly identified problem. Furthermore, some somatic variants may be selected for and gain widespread use. One such example is the semi-evergreen poplar, arising as a somatic mutation in the Lombardy poplar clone before 1900 (Pryor and Willing 1965).

Finally, as valuable clones are widely deployed, it is likely that they will encounter a variety of ecological conditions. Early testing typically includes only one or a few sites, and thus the clones' ranges of adaptation are frequently not known until later in a clonal testing program (Libby 1987b; Lindgren, Chap. 4; Frampton and Foster, Chap. 9, both Vol. 1). One of the pleasant surprises in several clonal forestry programs is the wide adaptation exhibited by outstanding clones. For example, several poplar clones perform well in parts of Europe, New Zealand, and both North and South America, on a wide range of sites. One explanation may be the high level of heterozygosity of these clones. Such broad adaptability is a useful characteristic, and can have great economic importance. Recent concerns about climatic stability make the strategy of selecting broadly adapted clones appear better than attempting to correctly deploy highly interactive clones to their best sites, as the climate may change on such sites during the rotation period of the deployed plantations.

2.3.2 Economic Concerns

Generally speaking, economic restrictions are less stringent than are biological restrictions. Economic restrictions can often be reduced or removed by solving technical problems such as large-scale vegetative propagation, or by silvicultural practices such as modifying spacing in response to opportunities provided by the greater reliability of clones.

Seedling plants are usually cheaper than vegetative propagules. When this constraint exists, it makes sense to propagate vegetatively only if the clonal material has greater value than the available seedling material. If the differences in values are large, relatively high prices per clonal propagule can be justified.

As discussed earlier, the risk of failure or unacceptable loss is greater if the rotation time is long and testing has been done for only a part of the rotation period. This is true of both seedlings and vegetative propagules, and it is a problem that includes biological, economic, technical, and ethical components. With fast-growing clones, the rotation time may be shortened, and thus the risk may diminish. There is a tendency to reduce rotation time even with seedling plantations, both to reduce risk and to more quickly recover the plantation costs. With greater initial costs and greater perceived risks, clonal forestry has logically developed in short-rotation situations. Programs with poplars, willows, and eucalyptus are examples (Zsuffa et al., Chap. 5; Zobel, Chap. 7, both Vol. 2). Deployment of clones in long-rotation forestry has been less common, with sugi (Ohba, Chap. 4, Vol. 2) serving as a rare example. Practical programs are now beginning with spruce, anticipating rotation times of 100–120 years; these are at present among the longest rotations contemplated. Oaks, with rotation times of about 200 years, are being clonally propagated for only specialty purposes at present.

The numbers of clones maintained and deployed have fiscal consequences. It is comparatively inexpensive to start a clonal program for a minor species with a limited plantation area. *Prunus avium*, for example, has been planted with a limited number of clones on specific sites with good economic results in France, and similar practice is now beginning in Germany. Since this species is often planted in mixed-species stands, there is little risk in such a program. By contrast, a species like Norway spruce is planted over a large area in monospecies plantations. A much broader genetic base is needed for such a program, and a transitional phase mixing appropriate seedlings with clonal propagules, either as Widespread Intimately Mixed Plantations (WIMPs) or Mosaics of Monoclonal Stands (MOMS), seems both biologically wise and economically prudent. *Prunus avium* plantations of about 100 clones are used in WIMP deployment in France, while several thousand clones of Norway spruce typically are deployed in WIMPs in Lower Saxony (FRG). However, a wide range of practices exists even for a single species such as Norway spruce, depending on local opinions, experience, and relevant laws or regulations (see Lindgren, Chap. 4, Vol. 1; Bentzer, Chap. 6; Muhs, Chap. 13, both Vol. 2). As a general principle, higher numbers of clones are generally deployed during early stages of clonal programs; as testing and experience develop, the numbers of deployed clones are generally reduced. Interestingly, the opposite trend often occurs with poplars, where numbers of deployed clones increase as the number of approved available clones is increased. With many poplar programs, the need for extensive testing of a clone prior to approval is considered more important than the need for diversity in production plantations.

2.3.3 Technical Concerns

Most technical restrictions are transitional, and are reduced as experience is gained and techniques improve. This is especially true of propagation techniques. The rapid development of various propagation techniques during the last few decades indicates that even today's difficult-to-handle species may be easy to propagate tomorrow. The fields of tissue culture, cell culture, and somatic embryogenesis are particularly promising, leading to the expectation that major improvements in vegetative propagation techniques will continue to become available for some time.

Techniques for effectively mixing propagules for a WIMP deployment are becoming available (see Appendix in Fasoulas 1988). The techniques for rapid and automatic board sorting in a lumber mill and of bottle handling in a soft-drink plant provide insights to the future.

Theoretically, the allocation of specific interactive clones to their best sites would improve overall productivity. However, the testing of clones and the level of site characterization required for such specific allocation are technically daunting. Furthermore, the suspicion that the climate and therefore the site may change during the rotation makes this option unlikely to be commonly used.

Regulatory and administrative restrictions can create technical difficulties that are serious. This is particularly true when these differ from place to place, so that general solutions and rules cannot be developed. Laws and regulations developed in support of changing forest values will require attention and perhaps technical innovation. These hopefully will be developed with meaningful input from practicing foresters. There is a danger that some laws or regulations might be developed as a way of impeding clonal forestry, by well-meaning people who see clonal forestry as unnatural. This is a significant problem, and will require educational and political rather than technical solutions.

It is comparatively easy to identify a particular genotype out of a set containing only a few well-known clones. However, as the number of clones contained in a mixture increases, so does the difficulty and expense of verifying identity. Traditional and modern methods of verifying clonal identity are presented in Cheliak, Chap. 8, Vol. 1.

2.3.4 Ethical Concerns

There are at least two main sources of ethical concern. There is increasing concern worldwide about the manipulation of nature. Clonal propagation is, or at least appears to be, more manipulative of nature than is seedling forestry. Furthermore, it opens the door for even more manipulation such as genetic engineering (see Chaps. 10–15, Vol. 1). Forests are often the last available ecological niches for native animals and plants. They are expected to fulfill many functions in addition to the renewable production of wood. In many forests, watershed management, protection from erosion or avalanche, recreation, or refuge for wildlife are determined to be more important than wood production.

Particularly with respect to recreation and wildlife habitat, plantations lose part of this function, and monoclonal plantations lose even more. Clonal forestry can be adapted to serve each of these additional functions, arguably better than can plantations of seedlings. But some forests should be left in their natural states, as self-regulating ecosystems, and clonal forestry has no place in these. The solution requires a recognition by all parties that not all forests are the same, nor do they all have the same purposes. In general, clonal forestry will be more at home where management is more intense, and, in many cases, clonal forestry will move forestry further from natural systems and closer to agricultural practices. Its proper place in the larger scheme of things needs to be considered as programs move from conventional to clonal forestry. As noted above, widespread use of the clonal option has severe implications for the natural gene pool of the species being cloned, as well as for the gene pools of those wild species that interact with the cloned species. Even if very large numbers of clones are maintained in the breeding lines, there will be considerable genetic losses. Because of the awareness that mistakes can be made, or that human curiosity will wish to investigate the pre-domestication genetic architectures and conditions of important domesticated plants (Zohary and Hopf 1988), we expect and predict that most or all serious clonal forestry programs will support and even include a significant component of genetic conservation (Ahuja, Chap. 10, Vol. 1; Millar, Chap. 3, Vol. 2).

3 Applications in Production Forestry

In this section, we focus on forestry in which a principal objective is the production of wood as a renewable resource.

3.1 Past Applications

Foresters and arborists have used vegetative propagation over a long period of time. Many of these early uses involved amenity and shelter plantings in rural settings, and then landscape plantings in urban settings as cities developed and grew. Sugi has been propagated as rooted branch cuttings for production forestry purposes since about 1400 in Japan (Ohba, Chap. 4, this Vol. 2), and poplars and willows have been vegetatively propagated, frequently by planting unrooted cuttings, for 3–5 centuries in Europe, Asia and the Middle East (Zsuffa et al., Chap. 5, Vol. 2). Air layering of fruit trees using containers with soil in the crowns of the trees was described by PLINIUS (76), and certain *Ficus* clones may have been serially grafted over a period of about 3000 years (Curtis 1968). Methods of vegetative propagation were also described in the Bible (Ezekiel 17:22–24).

Beginning in the middle of this century, increases in plantation forestry and the development of tree-improvement programs resulted in much interest and research in both sexual and vegetative propagation. At first, most angiosperm trees and all but sugi, junipers, yews, and *Chamaecyparis* among conifers were regarded as difficult to root. Many of the early attempts at vegetative propagation were motivated by a desire for uniform material for research (see Stettler and Ceulemans, Chap. 6, Vol. 1), or were in support of specialty purposes such as breeding-orchards or seed-orchards within classical seedling-based tree-improvement programs. But, as problems with vegetative propagation yielded to research and development, the deployment of clones for various practical forestry purposes became an interesting and obtainable possibility.

The development of clonal plantations of poplar, in the northeastern United States, in Italy and adjacent Mediterranean countries, in Belgium, the Netherlands, Germany, Australia, and elsewhere, provide the best examples of not just the use of clones, but of the development of full clonal forestry (see Libby and Ahuja, Chap. 1; Zsuffa et al., Chap. 5, both Vol. 2). Some of these plantations have proven records of outstanding productivity. But these early pioneers did not always find the right solution, and we can learn from their mistakes as much as or more than we can learn from their successes.

The first Norway spruce stecklings produced by human intervention were cuttings rooted in 1828 by Pfifferling (1830). These results were without immediate practical consequences. It was not until 100 years later that methods of vegetative propagation were actively developed, to serve the now classical approach to forest-tree breeding. This was most often done with the intention of producing clones for seed-orchards, rather than for direct clonal deployment to plantations.

Gardner (1929) demonstrated that cuttings from young donors could be rooted much more easily than could cuttings from older donors. Since the aim was to root from trees selected at or near harvest age (Snow 1938), Gardner's results were of little apparent help. With the detection and synthesis of phytohormones in the 1930s, numerous experiments for rooting spruce cuttings were initiated (Grace 1937; Thimann and Delisle 1939; Griffith 1940). Optimal timing, treatment, phytohormone dose, and effects of donor age were described. These researches continued to focus on propagation from older trees by improvement of rooting technology (Deuber 1940; Deuber and Farrar 1940).

In Germany, work with Norway spruce cutting-propagation was started in 1948 by R. Kleinschmit (Heitmüller 1951; Fröhlich 1955, 1957, 1959, 1961; Kleinschmit and Fröhlich 1956; Kleinschmit 1957, 1958, 1961). At about the same time, Norway spruce cutting propagation was started in Norway by Ruden (1965). Even then, the donors were usually more than 10 years old, often more than 20 years old, and efforts were still directed towards rooting from trees as old as possible. The concepts were reconsidered as it became obvious that old spruce trees are very difficult to root, and that the successful stecklings exhibit plagiotropism for a long time. This resulted in programs with early selection of vigorous young plants from tested provenances or progenies. It was then logical to deploy these clonally to production plantations, continuing to test and cull the clones beyond the early selection steps (Kleinschmit et al. 1973).

This conceptual change initiated a new wave of activities in cutting propagation of spruce in Europe, spreading in Germany and to Denmark, Finland, and Sweden, and later to France, Belgium, and Great Britain. These activities are summarized in three publications (Swedish, 1976; Swedish, 1977, 1981). These recent clonal programs are all closely linked to breeding programs, and they are developing into more or less large-scale operational programs (Bentzer, Chap. 6, Vol. 2).

3.2 Present Applications

Present applications of clonal forestry are largely covered in Chapters 4 through 12, Vol. 2, with some special topics appearing in Sections 4, 5, and 6 of this Chapter. Uses of clones include such diverse applications as research, vegetative multiplication, genetic conservation, deployment of specialty clones, and, not covered in any of the above, application of full clonal forestry in the c-p seed-orchard that is a cornerstone of New Zealand's family-forestry program (Libby 1990).

The focus of this volume is present applications in production forestry, and, rather than summarize them here, the reader is referred to the examples of full clonal forestry presented in Zsuffa et al. (Chap. 5), Zobel (Chap. 7) and Strobl and Evers (Chap. 12) and to the examples of programs still moving toward full clonal forestry in Chapters 4 (Ohba), 6 (Bentzer), 8 (Gleed), 9 (Rydelius and Libby), 10 (Mascarenhas and Muralidharan) and 11 (Russell).

3.3 Future Applications

On the basis of the present situation, described above and in other chapters of this Volume, it is possible to extrapolate some trends.

In the future, vegetative propagules will play an important role in afforestation and reforestation in all countries. Progress towards clonal forestry will be much more rapid in intensively managed short-rotation plantations than in long-rotation forestry.

It will be necessary to develop administrative rules that will guarantee that some of the various problems that can occur with clonal material are minimized, while encouraging and even protecting the positive aspects of clonal forestry.

Maintenance of genetic variation and diversity in the species being cloned and in their associated ecosystems will be a normal and high-priority part of most clonal forestry programs. The level of attention that needs to be given to maintenance of genetic diversity is influenced by several factors, including extent of clonal deployment, average use per clone, testing time, degree of environmental heterogeneity, and rotation time. These influences on decisions affecting genetic variation are presented diagramatically in Fig. 2.

With respect to diversity and the need for active genetic conservation, both depend in part on the stage of progress of the clonal program. Generally, as more is known about the tested clones, fewer clones can and probably will be deployed. While fewer clones are deployed as a program advances, this does not necessarily mean that either numbers of genotypes or their diversity will decrease within the continuing breeding population(s), particularly if sublining or multilining is practiced. However, these advanced generations will increasingly depart from their ancestral populations, and also from advanced genera-

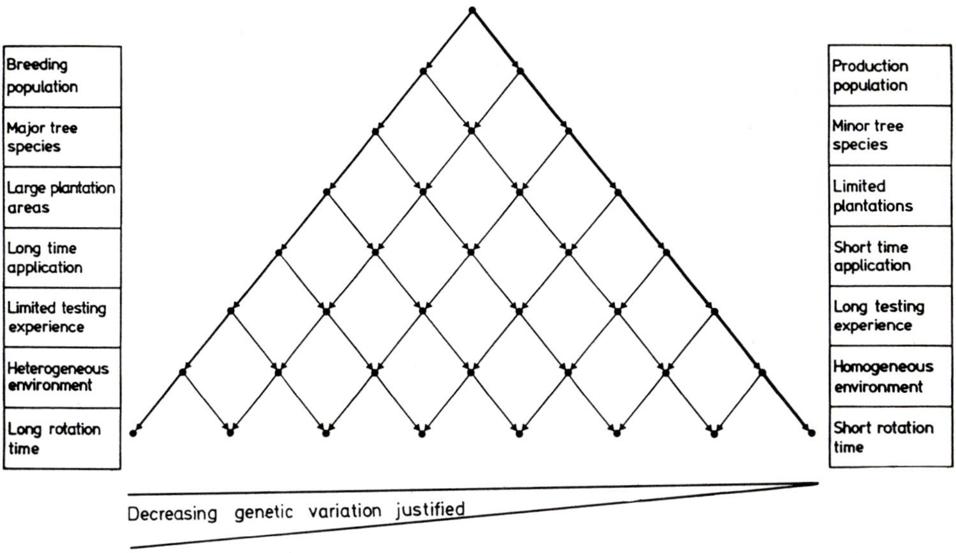

Fig. 2. Influences on decisions concerning genetic variation (clone numbers)

tions of such ancestral populations had evolution been allowed to progress without human influence.

Problems with long-term deployment of outstanding clones may arise due to the maturation of those clones. For example, with Norway spruce, six or seven cycles of propagation are clearly possible; but there are indications, with both serial propagation and hedging, that maturation may set limits to the propagation of even the very best selected clones. At least two courses of action seem possible.

1. The genes of the best clones may be forwarded in the generative cycle by controlled crosses among them. Outstanding clones can be selected from the progenies. The disadvantages of this option include the loss of the specific gene combinations of the parent clones, and the need to discontinue parent clones that have a wealth of accumulated test data and management experience with them.

2. Methods of maintenance of juvenility or of rejuvenation may be improved. Here the prospects for progress are promising (Bonga and von Aderkas, Chap. 12, Vol. 1). For example, sites near meiosis in flowers appear to be favorable sources of explants (Mullins and Srinivasan 1976; Eichholtz 1979; Jörgensen 1989) that may produce rejuvenated (or at least juvenile) tissue that can then be propagated as a juvenile form of the clone (Meier-Dinkel and Kleinschmit 1989). The storage of such somatic embryos is a technique for juvenility maintenance now under active investigation at Petawawa (Ontario, Canada) and elsewhere. Progress in flower induction makes it possible to reach flowering in Norway spruce and other conifers before age 10. This means that flower tissue is available for rejuvenation of clones at an age before maturation

of the ortet and of derived donor ramets is sufficient to seriously affect the production-level stecklings deployed from them. This second option, that of periodically replacing the current cutting donors with rejuvenated donors, or with stored donors whose juvenility has been maintained, has the advantage that exactly the same tested and increasingly known genotypes can be reproduced for long periods of time.

A combination of both of these options would be useful, with the first serving both clonal and zygotic forestry, and the second allowing the development of full clonal forestry.

Experience with other plants and animals being brought under domestication indicates that managers will opt for the greater uniformity, greater reliability, and higher productivity available with well-tested varieties and clones. While there are important differences between agriculture and forestry, the similarities are sufficient so that such a trajectory, once begun, seems likely to continue; and, as with well-known agronomic varieties, it seems likely that the change to known and reliable clones of forest trees will occur first where the economic values are the highest, and the sites the most productive.

Trends in deployment options vary. In some regions, widespread intimate mixtures of many clones (WIMPs) may be favored, or even required by regulation (Muhs, Chap. 13, Vol. 2). Mosaics of monoclonal stands (MOMS) may be favored in other regions and, in the absence of prohibiting regulation, this seems to be a common option as clonal forestry knowledge develops (Libby 1987a; Zobel, Chap. 7, Vol. 2). Among the considerations driving this choice are the substantial advantages of managing the MOMS configuration (Zobel and Talbert 1984, Chap. 10). However, a significant management risk associated with a MOMS configuration is the tendency for managers to concentrate on too few clones. Such cases have been noted with poplar MOMS (Guzina 1981) and in some regions of *Cryptomeria* deployment [pers. commun., R. Toda to JK 1973], where the single best clone in each specific region was always favored in that region. Finally, when a great deal is known about sets of clones, including data on their biotic interactions, their silvicultural requirements, and their likely end products, then specifically sequenced WIMPs may become the configuration of choice.

4 Applications in Urban Forestry

Landscape trees produced by vegetative propagation have been planted in rural and urban settings for centuries (Li 1963). Clones of weeping willow, Lombardy poplar, and London plane are the oldest best-known such trees, each having originated over 300 years ago.

Trees produced by budding or grafting on genetically variable seedling rootstocks usually exhibit aboveground clonal uniformity that is similar to that of clones grown from rooted cuttings. We will use the convention of loosely referring to all vegetatively propagated plants from the same ortet as a clone in this section, including those whose roots are not their own. Please also note that, while interest in urban trees is worldwide, most of the information in this section pertains to Europe and North America.

Why have clones become so popular for urban planting? Urban plantings have several qualities favorable for clonal forestry: the large investment made in each planted tree; the very high cost if it fails; and the high cost of having a tree misbehave, such as growing too large in the wrong place. Therefore, the performance superiority, trait specificity, uniformity, and reliability that can be achieved by using well-known clones is especially valuable. How are these clones developed and produced? What are their advantages and shortcomings? These considerations are interesting in themselves, and may also be useful in relation to the expanding uses of clones for wood, fiber, and other purposes.

4.1 Urban Tree-Planting Practices

Trees are planted in towns and cities for two main purposes: beauty and shade. Their functions are both esthetic and utilitarian, by making us more comfortable during hot weather and by making our surroundings more attractive, thus enhancing property values, enticing shoppers downtown, and providing other kinds of benefits as well. Some secondary uses include screening against sound, wind, and visual intrusion, for example.

Landscape trees are often highly valued as individuals, in contrast to forests where stands of trees are managed as populations. Desirable traits include healthy foliage, beautiful flowers or fruits, and special crown shapes. A diverse assortment of sizes, forms, textures, and colors is needed by those who creatively place trees in city scapes.

The art and science of growing trees in public places and along thoroughfares, around residences, and on commercial properties, have their roots in landscape gardening. Combined elements of horticulture, arboriculture, landscape architecture, and forestry have only recently been dubbed "urban forestry". This newer concept takes a more comprehensive view of the public and private management of trees by various agencies, organizations, and individuals within a municipality. Among those who work with urban trees are municipal and consulting arborists, tree commissions, park superintendents, landscape contractors, commercial groundskeepers, and homeowners.

Urban site conditions differ from natural environments to which native trees are adapted. Parks and spacious grounds may be similar to natural habitats, but street trees in planting strips or containers are often severely constrained, both above- and belowground. They are commonly hemmed in by buildings, overhead wires, and underground utilities, and constrained by the clearances required for vehicles, pedestrians, signs, and street lights. Urban trees encounter mild to severe stresses caused by compacted soil, drought, overwatering, heat, road salts, noxious fumes, lawn chemicals, and other unusual conditions that are imposed by their urban sites. These can and do have adverse effects on both native and exotic species. Some of the species that can best tolerate such stresses have previously adapted in nature to sites that experience both flooding and drought (Ware 1989).

The characteristics for which species and cultivars are selected for planting are highly site-specific. They are determined by the landscape effect that is to be created in a particular place, the stresses that trees will encounter there, space limitations, and the maintenance needs that may be anticipated. The reliable and predictable performance of clones therefore makes them especially desirable for many situations in towns and cities. A survey of 93 municipal arborists in the northeastern USA (Gerhold and Steiner 1976) found that they use many traits in selecting species and cultivars, involving their appearance, suitability for site conditions, and maintenance considerations (Table 2). The survey also found that some traits, such as pollution resistance, salt tolerance, and reduced maintenance requirements, would be used more commonly if better information about them were available.

An extensive array of vegetatively propagated landscape-tree cultivars has been produced (Table 3), although only a fraction of these are commonly deployed. The list of 941 cultivars by Santamour and McArdle (1982a, b, c, d, 1983a, b, c, 1984, 1985a, b, 1986), McArdle and Santamour (1984, 1987a, b) and Santamour et al. (1983) includes many of the species commonly planted as urban trees in the USA, but not all of them. For example, validated lists have not been published yet for *Crataegus*, *Malus*, *Prunus*, *Ulmus*, and *Zelkova*, which encompass a large number of urban-tree cultivars, nor have cultivars of conifer genera been so listed. It is noteworthy that many of the tabulated cultivars originated before 1900. However, in certain species (notably *Acer rubrum*, *Acer saccharum*, *Cornus florida*, *Cornus kousa*, *Fraxinus pennsylvanica*, *Gleditsia triacanthos*, *Pyrus calleryana*, and *Quercus palustris*), a majority of the cultivars originated after 1945, principally in North America. Many invalid cultivar

Table 2. Characteristics used by municipal arborists in northeastern USA for selecting species and cultivars (Gerhold and Steiner 1976)

	Percent of respondents who used the characteristic to select among:	
Characteristic	Species	Cultivars
Branching habit	97	79
Crown shape	94	76
Crown size	95	75
Trunk straightness	92	71
Disease resistance	98	68
Leaf size/color	88	67
Insect resistance	94	65
Fruit size/color	83	65
Flower color	80	63
Rooting habit	95	62
Drought tolerance	91	61
Cold tolerance	90	61
Salt tolerance	85	58
Air-pollution tolerance	73	49

Table 3. Number of validated cultivars in a partial list of hardwood tree species planted in urban landscapes in North America and Europe

Species	No. of cultivars	
	Total	New since 1945[c]
Acer platanoides L.	88[a]	23
A. rubrum L.	37	20
A. saccharinum L.	58[a]	8
A. saccharum Marshall	26	21
Cornus florida L.	60	52
C. kousa Hance	21[b]	21[b]
C. nuttallii Audubon	7	6
Fraxinus americana L.	21	9
F. angustifolia Vahl	24[a]	3
F. excelsior L.	91[a]	8
F. holotricha Koehne	1	1
F. latifolia Bentham	2	1
F. nigra Marshall	3	1
F. ornus L.	11[a]	3
F. pennsylvanica Marshall	20[b]	14[b]
F. uhdei (Wenzig) Lingelsheim	4	4
Ginkgo biloba L. (excluding oriental cultivars)	18	12
Gleditsia triacanthos L.	51	40
Liquidambar formosana Hance	1	1
L. styraciflua L.	21	15
Liriodendron tulpfera L.	22[a]	2
Pyrus calleryana Decaisne	10	10
Platanus species and hybrids	18	11
Quercus, subgenus Quercus		
Q. alba L.	5	0
Q. dentata Thumb.	1	0
Q. frainetto Ten.	1	1
Q. ilex L.	17	2
Q. macrocarpa Michx.	3	1
Q. petraea Mattuschka	35	4
Q. pubescens Willd.	6	0
Q. pyrenaica Willd.	1	0
Q. robur L.	119	6
Quercus hybrids	7	3
Quercus, other subgenera		
Q. acutissima Carruthers	1[b]	1[b]
Q. castaneaefolia Meyer	5	1
Q. cerris L.	7	0
Q. coccinea Muenchhausen	5	0
Q. glauca Thunb.	1	0
Q. × *hispanica* Lamarck	15	1

Table 3 continued

Species	No. of cultivars	
	Total	New since 1945[c]
Q. libani Oliver	1	0
Q. palustris Muenchausen	7	6
Q. phellos L.	4	1
Q. rubra L.	5	0
Q. shumardii Buckley	1	1
Q. velutina Lamarck	5	0
Tilia americana L.	8	4
T. × *europaea* L.	4	0
T. cordata Miller	18	16
T. flavescens A. Braun	3	1
T. platyphyllos Scopoli	27	5
T. tomentosa Moench	12	2
Tilia other hybrids	2	1
Total	941	343

[a] Many named before 1900.
[b] One seed-propagated cultivar.
[c] Date of origin is uncertain in some cases, and could be before 1945 despite later citations.

names were found by Santamour and coauthors, indicating that confused identities abound, whether accidental or intentional.

4.2 Production of Cultivars in Nurseries

Large wholesale nurseries typically produce a large assortment of species, but few cultivars per species. Most of the cultivars that are known are not commercially available except at speciality nurseries (for example, Hillier's Nursery, Winchester, England). Generally, three–ten cultivars are offered for the best-selling species; larger numbers presumably have been deemed unprofitable. About 350 landscape-tree cultivars are currently grown in quantities of 2000 or more each in the USA (Wandell 1989). Those being produced can be grown to appropriate sizes at reasonable costs, and marketed at a profit. Nearly all landscape-tree cultivars are clonal (Table 3, footnote b). However, in a few species, most landscape trees are grown from seed; comparative data are not available.

Cultural practices in the nursery can profoundly influence the development of new cultivars, which must perform well both in the nursery and in the city to succeed commercially. Budding in mid-summer is the most common method of propagation. All rootstocks are grown from seed, which is often of unknown or unrecorded origin. Rootstocks are typically grown 2–4 years before budding. The year following budding is a critical one. During this year the scion grows rapidly (in some cases > 2 m), branches are pruned from the rootstock, and staking for temporary support may be needed. In 2–4 years after budding the trees reach commonly planted sizes, generally 3–6-cm caliper (basal diameter) and 2–4m height. They may have been transplanted once or twice during this period to increase growing space and to concentrate the root system. Those grown for street trees will have their lower branches removed and their upper branches pruned to develop desired crown structure. The cultural practices in the nursery vary depending on species or cultivar, and on the desired end product.

Grafting works better than budding with a few species, but stock/scion incompatibility can be a problem with either method. Inappropriate rootstocks can aggravate this problem, such as red maple budded onto silver maple, a practice that temporarily increases growth but also the percentage of failures. Delayed graft incompatibility has caused the pin oak cultivars Crownright and Sovereign to be withdrawn from production. Similar problems with budded red maple provoked a switch from grafting to rooting of cuttings. Tissue culture has recently come into use in genera such as *Amelanchier* and *Betula*, with good results. It is generally assumed that a clone having its own roots is advantageous, but there are some reports of inferior root systems resulting from cutting propagation. Surprisingly, given the values at stake and the many decades of accumulated experience, this has not been thoroughly tested and evaluated. In some cases, such as English walnut cultivars grafted onto Hinds walnut rootstock in California, the roots of the (local) Hinds walnut handle pathogens and drought better than do English walnut roots.

Some nurserymen have observed that root systems may be strongly affected by which clone is budded onto them, and the root systems then exhibit easily recognizable scion-specific differences (W. Wandell, pers. commun. to HDG 1988). For example, Greenspire linden is unusually vigorous, a trait believed to be due in part to its effect on seedling root systems (W. Flemer III, pers. commun. to HDG 1984). Such advantageous stock/scion relationships are not well understood, but they may be important in the development of some cultivars.

4.3 Genetically Improved Cultivars

Early genetic inputs to the developing profession of planting trees in towns and cities consisted simply of selection and propagation. Many of the older selections were made by plantsmen with keen eyes for the unusual, or even bizarre. Latin names were used to indicate the appearance of leaves, fruits, or growth habit; for example, cultivars of *Acer platinoides* include Albo-marginatum, Brevilobum, Columnare, Dissectum, Globosum, Macrocarpum, Microphyllum, Monstrosum, Pendulum, Pubescens, Pygmaeum, Quadricolor and Umbraculiferum. Few of these were propagated in large numbers, but they served rather as curiosities in estates and botanical gardens. Some older cultivars did become popular, and even became progenitors of newer cultivars. Schwedler Norway maple, which has reddish leaves in the spring, gave rise to the longer-lasting, deeper colors of Crimson King, Faasen's Black Royal Crimson and Royal Red.

After the end of World War II in 1945, as attention was again turned to goals of comfort and beauty, cultivars of a different kind became prominent. This occurred principally through the efforts of several nurserymen. Their "keen eyes" discerned more practical traits; they also understood principles of how trees can be selected for adaptation to urban sites, and they had a knack for adroit marketing. The improvement strategy was generally single-phenotype selection, in a few cases following purposeful controlled pollination and selection among the offspring, followed by clonal propagation. The important change was that selection criteria and procedures were oriented more toward urban conditions and needs. Dramatic gains were realized in a single cycle. Examples include honeylocust cultivars selected for vigorous growth, straight trunks, good branching habit, and freedom from thorns and seed pods; littleleaf lindens with compact crowns, vigorous growth, and leathery leaves (observed to be resistant to urban stress); and Callery pears having desirable crown form and branching, fruitlessness, and resistance to diseases, insects, and cold. Greater attention was also directed toward cultivars of several species that had narrower crowns suited to confined spaces, and smaller flowering trees that would not grow into electric wires and whose upright branching would nevertheless be above the heights of pedestrians and vehicles.

Scientifically designed breeding methods have only recently been applied in developing landscape-tree cultivars (Gerhold et al. 1982). A few tree-breeders in universities, arboreta, and government agencies have used breeding strategies in research and in practical improvement programs that include both trait-index and family-index selection and formal testing protocols. Selection techniques have been developed to screen candidates for traits such as resistance to Dutch elm disease, air pollutants, iron deficiency chlorosis, and climatic extremes. Controlled crosses within and among species have increased the variability available and have created new combinations of traits, leading to better opportunities for selection. Trees with desirable traits have been selected more effectively when grown under more controlled environments, or with greater statistical control in varying environments. More thorough methods for testing trees exposed to street-side conditions have been developed, as this has been perhaps the weakest link in the improvement process. With these advanced techniques, a supply of even better cultivars may be anticipated in the years ahead.

Will the planting of clonal cultivars in cities lead to calamitous losses from diseases, insects, and unanticipated environmental events? Critics fear that this could happen if some clones later turn out to be highly susceptible. Such clones could indeed be lost, as their ortets probably would have been, but that does not mean that clones should not be used. Indeed, tree populations in cities have become genetically more diverse as more improved cultivars of many species have become available. For example, the risk of catastrophic loss due to Dutch elm disease, which largely affected seedlings of a too-popular species, is now lower, as many clonal cultivars of many other species have been more commonly planted; furthermore, the smaller localized outbreaks that continue to occur can be treated more effectively. The proper deployment of diversified clones can thus reduce such risks and losses, whereas reliance on too few species can be highly risky even if no clones are used.

4.4 The Present and the Future

Proprietary rights of the owners of cultivars, just becoming issues in clonal forestry (Muhs, Chap. 13, Vol. 2), have been important for four decades in the development and production of new varieties. However, attitudes and legal protection differ greatly between countries, for example between the USA and The Netherlands (Gerhold 1986). US plant patents protect owners against infringement when a new variety is initially tried out and while investments are made for increasing production and for marketing, i.e., during the years before these costs can be recovered and profits realized. Such patents also enable owners to license trustworthy reputable nurseries to produce cultivars in return for royalty payments. This also gives some assurance that the identity of a cultivar, and thus its reputation, will not be compromised by irresponsible unmonitored propagation. In The Netherlands, plant patent rights are not recognized. Instead, a government agency inspects nursery stock identities and good health prior to sales, and also provides "genuine" propagation materials to preserve tree identities. Each of these two contrasting systems is regarded as successful by those who use it. Compared to the Dutch system, the US system provides greater incentives for developing new cultivars, but has limitations for exploiting foreign markets.

Nearly all of the popular urban-tree cultivars now in production have come from the private sector, most having been developed through simple improvement procedures. The procedures have been cost-effective for originators, in part because testing in cities has been done largely at the expense of customers. A complete accounting of all development costs perhaps could show greater genetic gains resulting from more advanced techniques at lower cost per unit.

Most US patent owners are nurserymen, and are thus mostly concerned about traits favorable for nursery production and successful marketing. While traits related to good health, longevity, and low-cost maintenance are also recognized as desirable, there is less incentive to develop cultivars with these traits because a longer time is required to evaluate them and purchasers are often more concerned with low prices than with long-term values.

Clonal cultivars that are currently in use have proven to be advantageous because they exhibit predictable, uniform superiority over the average of seedling populations; they can be produced in nurseries at reasonable costs, given their values; and, in some countries, proprietary rights enable the recovery of development costs with reasonable expectation of profits.

To achieve cultivars with the full range of desired traits for both nursery and urban performance may require closer cooperation among researchers, breeders, producers, and users of urban trees. Cooperative improvement programs similar to those focusing on forest trees could be developed. The organization of such cooperative programs is complicated and yet it is crucial to their success. In setting priorities among species and traits, consideration must be given to the assortments of cultivars currently in production and the vested interests in those cultivars. Selection would often be based on evaluation of several traits both in the nursery and in a range of urban environments. In some cases, as a successful new cultivar is phased in, one of the poorer cultivars then in production will be phased out. Thus, release of new cultivars will affect and should be integrated with nursery production schedules and marketing strategies.

Despite such difficulties, cooperative urban tree-improvement programs should be feasible if incentives for commercial investors are safeguarded. Their success will require integrating the interests and abilities of nurserymen, scientists and consumers in commercially oriented production systems (at least in countries that depend on private enterprise). They should devise ways of giving exclusive rights for each member to select clones and to patent the best of them. As plant materials move through the later phases of the improvement process, e.g., from breeding to candidacy to testing to selection, records must become confidential and ownership of clones must be established and then protected. In this way, assurance can be given to profitability of new cultivars, which is the incentive for commercial support of urban-tree improvement programs.

5 Applications in Agroforestry

Agroforestry is a recent term for an age-old practice of intercropping, combining agriculture, horticulture, animal husbandry and forestry such that trees being grown for food, fodder and/or forest products are used in mixed or mosaic plantings with agricultural crops and animals (see Khosla and Khurana 1987 for details of many agroforestry programs). It has two important features: sustainability and polyculture (A.M. Schultz, as presented to West. For. Genet. Assoc. 1988 Meeting, Davis, Ca., in an address: *New Uses of Genetics for Agroforestry*).

Agroforestry has generally been associated with small farmers who often wanted or needed to be self-sufficient; hence the growing of a variety of crops, animals, and trees in close or intimate mixtures. As large-scale farming developed, the planting of single crops became common practice. However, some of these large monocultures are now recognized as being economically or ecologically detrimental, particularly in tropical conditions. One solution is to design agroforestry mixes that reduce the detrimental effects of cropping and perhaps even enhance the site in terms of nutrients and soil characteristics, while optimizing use of the space available. One may also pick quite different combinations of traits when breeding for harvest indexes in agroforestry settings involving small-scale operations and multiple products (including shade, N-fixation and erosion control), compared to harvest indexes for mass-production of few products (Schultz, cited above). Thus, interest in agroforestry has recently increased, and research investigating existing systems and creating new ones is being done.

5.1 The Past and Present Role of Clones

Cloned trees have been used for many years in agroforestry, employing clones of bananas, of stone-fruit trees, and of such genera as *Populus* and *Salix*. In general, these were species that were easily cloned, or whose value as clones had been proved and developed in other settings and adopted in agroforestry.

A contrast of agroforestry and urban forestry may be useful in order to understand why, even though both systems have been practised for many centuries, urban forestry has used clones much more extensively than has agroforestry. This difference may be best explained by the cultures that gave rise to these two types of forestry. Urban forestry has its roots in amenity horticulture, largely a phenomenon of developed countries. These countries have a long history of their people traveling around the world collecting plants of utilitarian/esthetic value. Other people were prepared to pay large sums of money for rare and interesting plants, and thus vegetative propagation of these plants was developed and came into common practice. In contrast, agroforestry has its roots in small subsistence farming, and long-distance contact among subsistence farmers and the resulting exchange of valuable plant material is likely to have been minimal. During the flowering of urban forestry in the middle of this century, large monoculture farms were replacing small agroforestry farms, and thus little attention or funding was directed to development of clones specifically for agroforestry until recently.

Several features of agroforestry currently make cloning of trees attractive. These include the importance of the small number of trees normally included in a subsistence unit to the sustainability of that unit, even if their absolute values are not high. Some traits are unusual, and specific for intimate mixtures with other crops. These include the form of the root system, preferably being deep-rooted to avoid near-surface competition with annual crops; and the form and density of the crown, preferably being narrow-crowned or of light density to allow sufficient sunlight to penetrate to the lower-level crops. In subsistence agroforestry, multipurpose trees are desired that can be used for such things as fodder, fuelwood, biofertilizer, and finally timber at various times of the year and the rotation, and, in common with production forestry, trees with rapid growth rates, reliable high performance and value, and exhibiting above-average resistance to various damaging agents are useful. The ability of the overstory trees to protect understory crops from harsh elements of the weather and perhaps also from some pests is a special feature of the latter trait useful in intimate-mixture agroforestry. Similarly, domestic animals are better sheltered and less likely to develop epidemic diseases under scattered tree overstory, compared to a concentrating shelter in otherwise open pasture.

5.2 The Future Role of Clones

The sites of agroforestry practice in the foreseeable future are likely to be either highly productive or substantially degraded. In either case, it should be possible to select clones that take advantage of the highly productive sites, or tolerate the problems of the degraded sites, better than would populations of seedlings.

Among the genera and species currently being used in agroforestry are the following:

Fuel, biomass, and timber: acacias, *Casuarina*, eucalypts, pines, and poplars
Food: apples, bananas, coconuts, figs, mangoes, mulberrys, papayas, pears, *Prunus*, and walnuts
Multipurpose: *Leucaena*, *Morus*, and *Paulownia*

Some of these are already normally deployed as clones; in others, cloning research and practice is well started. Other as-yet rarely cultivated species may join the above list, and selection of outstanding clones will be a likely route of their domestication for agroforestry purposes.

Risk management by the deployment of genetic diversity is a challenging problem. This is already achieved by the strategy of mixed cropping on the species level. But within the tree component, particularly if cloning becomes common, there is the risk of too few clones being deployed. Where agroforestry is practised by many small farmers in a region, each is probably too small to deploy the seven or more clones of each species considered appropriate for managing risk (Lindgren, Chap. 4, Vol. 1), and each is independently likely to wish to plant the one best clone available from each species. Some regional cooperative organization may be necessary in order to equitably counter risk of epidemic because of too few clones, both for individual farmers and regionwide. In larger-scale agroforestry holdings, such as described below, such a risk-avoidance strategy can be adopted effectively by each unit.

An interesting simple agroforestry system is being developed in New Zealand and Australia (Reid and Wilson 1985). It melds ranching and production forestry, combining cloned radiata pine with selected range grasses to be grazed by selected sheep, cattle, and even red deer. Compared to seedlings, the use of cloned pines allows much wider initial spacing, thus greatly reducing establishment costs while maintaining productivity. The wider spacing, in turn, allows the space between the trees to be used better and longer by the grazing animals. Economic analyses of this system have found that normal or even increased returns on the timber trees are obtained, and additional returns on the animals become almost pure profit (Knowles and Percival 1983). Variations on this theme are being developed in other regions with other species, such as poplars, corn and cattle, or walnuts and grazing animals.

5.3 Breeding for Agroforestry

Breeding in support of either clonal or zygotic agroforestry programs is still primitive. Burley (1987) summarized and discussed some of the special problems associated with such breeding. These include the number of candidate species, perhaps over 2000; likely negative genetic correlations among traits in multipurpose trees, particularly among productivity for wood, fodder and fruit; genotype-by-environment interactions, not only among a range of sites from degraded to productive agricultural soils, but also including biotic interactions among the several components of the polyculture; and the uncertain history of many landraces currently in use. A recent example of the latter problem is the realization, brought on by pest susceptibility, that many landraces of *Leucaena* have narrow genetic bases, some perhaps originating from single parent trees (pers. commun. J. Brewbaker to WJL, 1990). Burley recommended the development of multilines as a response to the multiplicity of agroforestry objectives. The selection of particularly useful and contrasting clones from these multilines will serve such a variety of objectives well.

6 Applications in Christmas-Tree Production

Christmas-tree production has many of the qualities that make some of the advantages of clonal forestry attractive (see above, this chapter, and Libby and Ahuja, Chap. 1, Vol. 2). This is particularly true for the so-called choose-and-cut operations, where the buyers visit a Christmas-tree farm and choose among the trees growing on site.

There are five ways in which a grower can increase efficiency by improving the quality of planting stock: (1) by increasing the percentage of trees that are sold, i.e., by reducing the percentage of culls; (2) by reducing the labor and other management costs needed to bring a tree to harvest; (3) by decreasing the average time from planting to harvest; (4) by increasing uniformity, particularly with respect to time from planting to harvest; and (5) by selecting adapted genotypes from a desired species that on average is not adapted to the plantation conditions. Site preparation, weed and pest control, irrigation, fertilization, and control of the genetic quality of planting stock all contribute to manipulating these components of efficiency. One can serve all of these goals by selecting populations or families of seedlings. However, the levels of control offered by cloning and application of clonal forestry principles are attractive. The option of finding outstanding Christmas trees and then cloning them is beginning to receive serious attention among Christmas-tree growers.

6.1 A California Example

In a recent study of radiata pine (*Pinus radiata*), the most common choose-and-cut Christmas tree in California, several questions relevant to the effectiveness of genetic inputs were addressed (Libby et al. 1985). The study was conducted in a plantation of 980 pedigreed trees, in which origin population, membership in a family (open-pollinated families in some cases, full-sib families in others), and in some cases clonal identity, was known for each tree. It combined ranked preferences of cooperating growers and consumers with observations and measurements of various appropriate qualitative and quantitative traits, made and recorded by university personnel.

The difference between the average of the growers' evaluations and the average of the consumers' evaluations was small. However, there were substantial differences among the growers and among the consumers. Detailed comparisons of these judges' scores indicated that they were often responding to somewhat different criteria when rating the trees. For example, in one year's evaluation, 550 of the 980 trees were selected as being acceptable and in the top 30% by at least one judge, but only six of them were selected as being in the top 10% by all six judges. It was concluded that this was not methodological or experimental error, but that there is valid diversity of opinion as to what constitutes a good, or desired, Christmas trees. There was complete agreement on a large class of trees that none of the judges selected. Most of these, presumably, would be trees that the consumers would prefer not to buy, and the growers would not wish to spend time, space, and money on. These accounted for 44% of the trees in the test.

Linear multiple-regression analyses were performed, correlating the judges' scores to the traits measured and observed by university personnel on the same trees. Of the 13 traits measured and observed, nine were identified as being among the five most important, in various orders, for one or more of the judges. The average scores of the six judges and the scores of most of the individual judges were most highly correlated with three traits: height, number of whorls, and branches per whorl. The other ten traits added little to the predictive value of the equations.

For some traits, such as needle length, the judges were often selecting against trees exhibiting either extreme, thus preferring trees exhibiting some intermediate value within the range available. Such traits would not be identified as contributing strongly to the linear multiple-regression equations used in the study, but they are likely to be important in influencing the judgement of both growers and consumers in selecting Christmas trees. It is for traits such as these that clonal selection and deployment have large advantages over selection and deployment of zygotic lines or families bred for intermediate values. Most Christmas-tree species have genetically variable populations, and the breeding programs that produce trees of interest to Christmas-tree growers are likely to increase the level of genetic variability in their breeding lines compared to that in native populations. This can result in substantial segregation of genetic variation both above and below the desired intermediate values.

The most important single trait in these analyses accounted for less than 15% of the variation in each of the judges' scores, and in the sum of all of the judges' scores, and the best combinations of five traits accounted for less than 25%. It appears that the best linear combination of the traits measured by university personnel is a poor substitute for the integrative scores of the judges, and a "Christmas-tree index" based on these component traits would not be very effective.

Standard heritability analyses yielded an average narrow-sense heritability of about 14% for the judges' scores, with higher values in some of the population-samples. This allows the prediction that there would be a weak-to-moderate response to selection and breeding for Christmas-tree performance. In many cases, each observer's judgement as to the overall balance of several traits was probably more important than the selection of individuals or clones with the greatest or least expression of one or more component traits. The "judges' scores heritability", was for such balanced evaluations, often relying more on impressions than trait-by-trait analyses.

One factor that lowers the likely response to breeding is the valid disagreement among both growers and consumers as to what constitutes a desired Christmas tree. If, in a "pick-the-winner" breeding program, one were to mate trees with similar index scores, but on which two judges disagreed as to details of their merits, the offspring might on average combine levels of expression for several traits that neither judge liked.

An alternative scheme might identify different groups of selected trees, each focusing on a different balance of various traits. One could then create multilines and breed within them. This would maintain and even increase the diversity of good trees available to growers and consumers, amplifying the different constellations of desirable trait expressions more effectively than would a breeding program based only on an index of judges' scores.

Whether combined with or independent of such breeding programs, cloning provides a quick capture of outstanding combinations of desired characteristics, with the additional advantage of being able to respond effectively to diversity of opinion as to what constitutes a desired tree.

6.2 Advantages and Disadvantages

In addition to maintaining diversity for both biological and marketing purposes, clonal forestry provides several other advantages to Christmas-tree growers. (1) Christmas trees are a short-rotation high-value crop. Thus, relatively high added costs of propagules can probably be justified and rapidly recovered. (2) The numbers of clonal propagules needed per year seem likely to be of magnitude 10^3 to 10^5, rather than the 10^6 or more needed to justify a typical seed-orchard operation. Managing a hedge-orchard (or some other form of clone donors) can probably meet these modest yearly demands for numbers of propagules more cost-effectively than can the seed-orchard alternative, particularly if contrasting and predictable diversity among the propagules is desired. (3) The uniformity available from monoclonal blocks of the deployed clones allows each block to be specifically managed as a unit, generally being mostly or completely harvested in the same year. (4) As has been found with radiata pine in California, some clones do well in most areas where radiata pine is grown, while others are more limited as to their appropriate regions of deployment or to particular systems of management. Clonal forestry offers specific deployment of clones to appropriate growers as performance records accumulate.

One disadvantage, at least with radiata pine, is the change in tree form with maturation. Stecklings generally have fewer whorls and fewer branches per whorl than do their seedling ortets, and thus only a minority of clones chosen on promising ortet performance can be effectively used for Christmas trees. This points up the need for testing not only the clones, but the chosen propagule type (see Frampton and Foster, Chap. 9, Vol. 1). However, one of the important diseases of radiata pine Christmas trees is western gall rust, and resistance to this disease increases with increasing maturation state (Zagory and Libby 1985). Thus, families or populations that might support epidemics of this disease if deployed as seedlings might be managed at sub-epidemic levels if selected clones could be deployed.

6.3 The Present and the Future

In about 1975, a single unusual tree appeared among about 1000 hybrid seedlings of *Pinus attenuata* × *radiata* being grown as Christmas trees in the central Sierra Nevada of California. This unusual individual was cloned by rooting cuttings and named the Hyder Hybrid. It has been under test at its parent farm and at a few other locations for over 10 years, and as of 1989 is in routine large-scale production on the parent farm. It has maintained its juvenility over this period of time, and has an advantageous combination of characteristics making it useful as a Christmas tree, in addition to the growth rate and cold- and drought-tolerance typical of this hybrid cross. It is, to our knowledge, the first large-scale routine use of a clone for Christmas-tree purposes.

In Florida, in 1982, the B. K. Lee Christmas-tree Farm became the first, to our knowledge, to be 100% clonal (pers. commun., S. Kossuth to WJL, 1989). Starting in 1980, outstanding eastern redcedars (*Juniperus virginiana*) were located in nearby graveyards, and their open-pollinated seeds were collected and germinated. Based on single-copy (Level I) selection, 126 of these seedlings were given clonal numbers and became the source of all new redcedar plantings in the farm. Early selections were based on ideotypes emphasizing form and color. In subsequent years, selection emphases have shifted to resistance to serious pests [particularly cedar mites (*Tirsetaeus cupressi*), phomopsis (*Phomopsis juniperovora*) and cercospora blight (*Cercospora sequoiae*)], reducing the original 126 red cedar clones to 60 being deployed in 1990.

A similar clonal program was developed for pines at the Lee Farm, using stump-sprouts from cut Christmas trees to produce stecklings of spruce pine (*Pinus glabra*) and cuttings from lower branches of selected Virginia pine (*P. virginiana*). However, the customers prefer Choctawhatchee sand pine (*P. clausa* var *immuginata*). Here, interspecific grafts are used, selecting scions from good Christmas-tree ideotypes and grafting them onto slash pine (*P. elliottii*) rootstocks. This greatly accelerates the growth rates of the sand-pine scions and secures reliably good form. Both the redcedar and sand pine are typically harvested 3 years after planting, and the best clones are now in their fourth serial propagation without apparent loss of performance.

In California, as of 1989, two large growers and a few small growers are deploying a few tested clones in modest numbers. If they are successful in this, and particularly if the knowledge and experience that they gain solves or lowers concern about the maturation-related problems, then many other Christmas-tree growers will soon join the ranks of those practising clonal forestry with radiata pine. They may do so using the most modern of techniques. In one recent analysis (Dammann 1988), it was noted that "Many growers are turning from cross-breeding and selection done by lumber and pulp industries as the source of their trees, to the laboratory, where they hope tissue-culture or cell-culture techniques will produce trees that grow at a uniform rate to permit mass harvesting."

References

Abbott AJ, Atkin RK (eds) (1987) Improving vegetatively propagated crops. Academic Press, New York London, 416 pp

AFOCEL (1982) Meet In vitro propagation of forest tree species, Fontainebleau. AFOCEL, Nangis, Fr, 363 pp

Burley J (1987) Strategies for genetic improvement of agroforestry trees. In: Khosla PK, Khurana DK (eds) Agroforestry for rural needs. Azad Hind, Chandigarh, India, pp 253–265

Carson MJ (1986) Advantages of clonal forestry for *Pinus radiata* – real or imagined? NZ J For Sci 16:403–415

Collins WH (1976) Nursery growing practices as related to selection and production of municipal trees. USDA For Serv Tech Rep NE-22:167–178

Curtis HJ (1968) Das Altern. Die biologischen Vorgänge. Fischer, Stuttgart, 105 pp

Dammann T (1988) Cloning the perfect tree. San Francisco Chron 27 Nov 88, Sunday Punch. p 5

Deuber CG (1940) Vegetative propagation of conifers. Trans Conn Acad Arts Sci 34:1–83

Deuber CG, Farrar JL (1940) Vegetative propagation of Norway spruce. J For 38:578–585

Dunlap JM (1990) Genetic variation in natural populations of *Populus trichocarpa* from four river drainages in Washington. PhD Diss, Univ Washington, Seattle

Eichholtz DA (1979) Adventive embryony in apple. Hortic Sci 14:699–700

Fasoulas AC (1988) The honeycomb method of plant breeding. Altidjis, Thessaloniki, 167 pp

Fröhlich HJ (1955) Untersuchungen über die autovegetative Vermehrung unserer Holzarten nach Anwendung von Wuchstoffen. Diss Univ Hannover-Münden, 133 pp

Fröhlich HJ (1957) Technische Einrichtungen zur vegetativen Vermehrung von Waldbäumen. Gartenbauwissenschaft 22:288–296

Fröhlich HJ (1959) Grundlagen und Voraussetzungen der autovegetativen Vermehrung. Silv Genet 8:48–58

Fröhlich HJ (1961) Untersuchungen über das physiologische und morphologische Verhalten von Vegetativvermehrungen verschiedener Laub- und Nadelbaumarten. Allg Forst Jagdz 132:39–58

Gardner FE (1929) The relationship between tree age and rooting of cuttings. Proc Am Soc Hortic Sci 26:101–104

Gerhold HD (1986) Workshop on urban tree cultivars. Trip Report. USDA Off Int Coop Dev, Neth, 1–5, Sep, 13 pp

Gerhold HD, Steiner KC (1976) Selection practices of municipal arborists. USDA For Serv Tech Rep NE-22:159–166

Gerhold HD, Steiner KC, Cech FC, Karnosky D (1982) Genetic improvement and urban trees: a problem analysis for environmental forestry research. USDA For Serv NE/NA, 65 pp

Grace NH (1937) Physiologic curve of response to phytohormones by seeds, growing plants, cuttings and lower plant forms. Can J Res Sec C15:538–546

Griffith BG (1940) Effect of indolebutyric acid, indoleacetic acid, and alpha-napthalene acetic acid on rooting of cuttings of Douglas-fir and Sitka spruce. J For 38:496–501

Guzina V (1981) Danasnje stanje oplemenjivanja crnih topola kod NAS Sumarski Inst, Jastrebarsko, Radovi 44:103–117 (in Yugoslavian)

Harper JL (1977) Population biology of plants. Academic Press, New York London, 778 pp

Heitmüller HH (1951) Beiträge zur vegetativen Vermehrung der Waldbäume. Diss, Univ Hannover–Münden, 120 pp

IUFRO (1974) Proc IUFRO Meet Vegetative propagation, Rotorua, NZ. NZ J For Sci 4:120–158

IUFRO (1982) Meet Breeding strategies including multiclonal varieties, Sensenstein, FRG. Dep For Tree Breed, Lower Saxony For Res Inst Escherode, 238 pp

Jörgensen J (1989) Somatic embryogenesis in *Aesculus hippocastanum* L. by culture of filament callus. J Plant Physiol 135:240–241

Khosla PK, Khurana DK (1987) Agroforestry for rural needs. Indian Soc Tree Sci, Azad Hind, Chandigarh, India, 363 pp

Kleinschmit J, Müller W, Schmidt J, Racz J (1973) Entwicklung der Stecklingsvermehrung von Fichte (*Picea abies* Karst.) zur Praxisreife. Silv Genet 22:4–15

Kleinschmit R (1957) Stecklingsvermehrung in Wasserkultur. Silv Genet 6:152

Kleinschmit R (1958) Nadelholzstecklinge. Forst Holzwirtsch 13:347

Kleinschmit R (1961) Versuche mit Fichtenstecklingen für einen genetischen Test. Silv Genet 10:10

Kleinschmit R, Fröhlich HJ (1956) Stecklingsvermehrung in automatisch gesteuerter Wassenkultur. Forstarchiv 27:149

Knowles RL, Percival NS (1983) Combinations of *Pinus radiata* and pastoral agriculture on New Zealand hill country: Forestry productivity and economics. In: Hathaway DB (ed) Foothills for food and forest. Timber, Beaverton, Or (also Or State Univ School Agric Symp Ser 2), pp 203–218 (383 pp)

Leakey RRB (1987) Clonal forestry in the tropics – a review of developments, strategies and opportunities. Comm For Rev 66:61–75

Li H-L (1963) The origin and cultivation of shade and ornamental trees. Univ Penn Press, Philadelphia, 282 pp

Libby WJ (1982) What is a safe number of clones per plantation? In: Heybroek HM, Stephen BR, von Weissenberg K (eds) Resistance to diseases and pests in forest trees. Purdoc, Wageningen, pp 342–360 (503 pp)

Libby WJ (1987a) Testing and deployment of genetically engineered trees. In: Bonga JM, Durzan DJ (eds) Cell and tissue culture in forestry, vol 1, Chap. 11. Nijhoff Junk, The Hague, pp 167–197 (422 pp)

Libby WJ (1987b) Testing for clonal forestry. Annales Forestales 13/1-2:69–74

Libby WJ (1990) Advantages of clonal forestry revisited. In: Proc Joint Meet W. For Gen Assoc; and IUFRO Working Parties S2.02-05, – 06, – 12, 14. Olympia Wa, Aug 1990. Weyerhaeuser, Tacoma, Wa, Sec. 5.18, 13 pp

Libby WJ, Rauter RM (1984) Advantages of clonal forestry. For Chron 60(3):145–149

Libby WJ, Fanger-Vexler MA, Russell JH (1985) Variation among Monterey pines of known genetic origins. Cal Christmas Tree Growers Bull 124:6–10

Martin B (1987) Amélioration génétique des *Eucalyptus* tropicaux. Contribution majeure à la foresterie clonale. Thesis, Univ Paris Sud, Centre d'Orsay, 218 pp

Mathias PJ (1988) Micropropagation of the tropical hardwoods, *Khaya ivorensis* and *Nauclea diderrichii*. PhD Diss, Univ Nottingham, 312 pp

McArdle AJ, Santamour FS Jr (1984) Checklists of cultivars of European ash (*Fraxinus*) species. J Arboric 10:21–32

McArdle AJ, Santamour FS Jr (1987a) Cultivar checklist of white oak species (excl. *Quercus robur* L.). J Arboric 13:203–208

McArdle AJ, Santamour FS Jr (1987b) Cultivar checklist of *Quercus* (excl. subg. *Quercus*) J Arboric 13:250–256

Meier-Dinkel A, Kleinschmit J (1989) Aging in tree species: present knowledge. In: Adv Stud Inst Molecular basis of plant aging, Ribadesella, Spain, 25 June–8 July, 13 pp

Mullins MG, Srinivasan C (1976) Somatic embryos and plantlets from an ancient clone of the grapevine (cv. Cabernet Sauvignon) by apomixis in vitro. J Exp Bot 27(100):1022–1030

Pfifferling (1830) Erfahrungen über die Nachzucht der Fichte durch Steckreiser. N Jahrb Forstk 7:54–62

Plinius GG (76AD) Naturgeschichte (Translation of Christian Lebrecht Strack) (1853–1855) XVII Buch, Ab 13, Pkt 21, Bremen. Heransgegeben 1968, Wissenschaftliche Buchgesellschaft, Darmstadt

Pryor LD, Willing RR (1965) The development of poplar clones suited to low latitudes. Silv Genet 14:123–127

Reid R, Wilson G (1985) Agroforestry in Australia and New Zealand. Goddard & Dobson, Victoria, 223 pp

Ruden T (1965) Stecklingsvermehrung von Fichten. In: Vortragsr 2. Symp Industriellet Pflanzenbau, Wien, pp 133–137

Santamour FS Jr, McArdle AJ (1982a) Checklist of cultivated maples I. *Acer rubrum* L. J Arboric 8:110–112

Santamour FS Jr, McArdle AJ (1982b) Checklist of cultivated maples II. *Acer saccharum* Marshall J Arboric 8:164–167

Santamour FS Jr, McArdle AJ (1982c) Checklist of cultivated maples III. *Acer platanoides* L. J Arboric 8:241–246

Santamour FS Jr, McArdle AJ (1982d) Checklist of cultivated maples IV. *Acer saccharinum* L. J Arboric 8:277–280

Santamour FS Jr, McArdle AJ (1983a) Checklist of cultivars of Callery pear (*Pyrus calleryana*). J Arboric 9:114–116

Santamour FS Jr, McArdle AJ (1983b) Checklist of cultivars of honeylocust (*Gleditsia triacanthos* L.). J Arboric 9:248–252

Santamour FS Jr, McArdle AJ (1983c) Checklist of cultivars of North American ash (*Fraxinus*) species. J Arboric 9:271–276

Santamour FS Jr, McArdle AJ (1984) Cultivar checklist for *Liquidambar* and *Liriodendron*. J Arboric 10:309–312

Santamour FS Jr, McArdle AJ (1985a) Cultivar checklists of the large-bracted dogwoods: *Cornus florida*, *C. kousa*, and *C. nuttalli*. J Arboric 11:29–36

Santamour FS Jr, McArdle AJ (1985b) Checklists of cultivars of linden (*Tilia*) species. J Arboric 11:157–164

Santamour FS Jr, McArdle AJ (1986) Checklist of cultivated *Platanus* (plane tree). J Arboric 12:78–83

Santamour FS Jr, He S-A, McArdle AJ (1983) Checklist of cultivated ginkgo. J Arboric 9:88–92

Schreiner EJ (1938) Forest tree breeding techniques. J For 36:712–715

Schreiner EJ (1970) Genetics of eastern cottonwood. For Serv Res Pap USDA WO-11, 32 pp

Snow AG (1938) Use of indolebutyric acid to stimulate the rooting of dormant aspen cuttings. J For 36:582–587

Swedish (1976) Breeding Norway spruce. Dep For Genet, Swed Coll For, Stockholm Bogesund, 205 pp

Swedish (1977) Vegetative propagation of forest trees – physiology and practice. Inst For Improv; and Dep For Genet, Swedish Univ Agric Sci, Uppsala, 159 pp

Swedish (1981) Symposium on clonal forestry. Sver Lantbruksuniv, Dep For Genet Uppsala. Swed Univ Agric Sci Res Note 32:131 pp

Symposium (1976) Symp Juvenility in woody perennials, College Park/Berlin. Acta Hortic 59:1–317

Thimann KV, Delisle AL (1939) The vegetative propagation of difficult plants. J Arnold Arb 20:116–231

Wandell WN (1989) Handbook of landscape tree cultivars. East Prairie, Gladstone, Il, 313 pp

Ware GH (1989) Trees for restricted spaces. Metropol Tree Improv Alliance Proc 6:80–85

Zagory D, Libby WJ (1985) Maturation-related resistance of radiata pine to western gall rust. Phytopathology 75:1443–1447

Zobel B (1985) Vegetative propagation in *Eucalyptus*. In: Zsuffa L, Rauter RM, Yeatman CW (eds) Clonal forestry: its impact on tree improvement and future forests. Proc 19th Meet Can Tree Improv Assoc, pt 2, Toronto, pp 136–144 (235 pp)

Zobel B, Talbert J (1984) Applied forest tree improvement. John Wiley & Sons, New York, 505 pp

Zohary D, Hopf M (1988) Domestication of plants in the old world. Clarendon, Oxford, 249 pp

Zsuffa L, Rauter RM, Yeatman CW (eds) (1985) Clonal forestry: its impact on tree improvement and future forests. Proc 19th Meet Can Tree Improv Assoc, Pt 2, Toronto, 235 pp

3 Conservation of Germplasm in Forest Trees

C.I. Millar[1]

1 Introduction

Forest trees meet human needs and contribute to natural ecosystems in unique and diverse ways. As a source of fuel and fiber, they are important commodities. As dominant members of many natural ecosystems, they play keystone ecological roles in plant and animal communities. As habitat for other organisms, forests are a reservoir of great natural diversity. In themselves, trees contribute taxonomic diversity through a wealth of species, from rare to widespread and through great genetic diversity within species. Forests purify water and cleanse the atmosphere. And, to increasingly crowded urban denizens, forests provide valued opportunities for solitude and recreation.

Understandably, there are as many kinds of demands for conservation of diversity in forests as there are diverse values of forest resources. Although some forest-conservation issues seem to apply specifically to the species, community, or ecosystem levels, genetic concerns in fact pervade all levels. The actual approach to forest-tree conservation and the methods used jointly depend on available genetic-conservation techniques and on the ultimate goal for perpetuating the germplasm.

1.1 Threats to Forest-Tree Germplasm

Conservation usually is motivated by a threat to natural resources, either an immediate crisis or a perceived future challenge. Because of the diverse roles that trees play in human and wild environments, trees face more kinds of threats than most classes of organisms. Over-exploitation, land conversion, and deforestation have extirpated populations and brought about species extinction, especially in tropical regions (references in Wilson 1988). Global climate change threatens both rare and widespread tree species with mass population extirpation and possible species extinction, especially at temperate latitudes (Hansen et al. 1987; Peters 1990; Davis 1991).

Forest trees also are threatened in a way that is not problematic for most other wild species. Because trees are managed in or near their native forests, the potential for gene-pool disruption due to production forestry and horticulture is

[1] Institute of Forest Genetics, Pacific Southwest Research Station, US Department of Agriculture, Forest Service, Berkeley, CA 94701, USA

Clonal Forestry II, Conservation and Application
Ed. by M.R. Ahuja and W.J. Libby

high. From dysgenic to progenic harvest, from operational reforestation to tree improvement, from landscaping to restoration, trees are cut, grown, and replanted in ways that can contaminate, impoverish, or alter the natural genetic architecture of forests (Ledig 1986, 1988; Millar 1987; Millar and Libby 1989a, 1991).

The use of clones poses special threats to genetic diversity in restoration, tree improvement, and general forestry. With the ability to replicate exact genotypes comes the potential to reduce diversity in restored forests and plantations by planting many copies of one or a few genotypes. The risks associated with genetic monocultures have been widely recognized and discussed in agricultural (NAS 1972; Frankel and Hawkes 1975) and forestry contexts (FAO/UNEP 1975; Ledig 1986). Most modern forestry programs take active measures to guard against low diversity, although from a strict production standpoint, as few as seven clones per plantation have been recommended as safe (Libby 1981).

The level of genetic awareness is not yet as high in most programs for restoration or conservation of rare species as it is in production forestry (Millar and Libby 1991). The former activities to date have been based primarily on ecological and botanical but not genetic principles. Clonal propagation is commonly used, especially for noncommercial species. The incentive to plant many copies of a few clones may be high, either for ease in production, or because the number of remaining wild genotypes available is low. Increasingly, however, attention is being given to the genetic risks associated with clonal replication of a few genotypes in restoration and in programs for rare plant recovery (Ambrose 1987b; Holsinger and Falk 1991).

Despite these threats, the genetic wealth of many natural forests is still great. Forest trees are among the most genetically diverse organisms known; their evolutionary persistence may depend on this. Trees have evolved many life history traits that promote and maintain genetic variability at all levels under natural conditions. This variation is the raw material both for natural selection of adaptations to changing environments and for artificial selection of traits desired by humans. Unlike many agricultural crops, trees have only begun to be domesticated (Libby et al. 1969), so that the gene pools are rich and have not been exploited. It is this natural variation that is sought and protected by breeders, conservation biologists, poets, and backpackers, and by the many forest organisms that depend on trees.

1.2 Approaches to Conservation

An important challenge to conservation is to conserve genetic diversity in ways that protect the germplasm in its desired form. Conservation strategies range from preservation of germplasm in cloned DNA sequences and genomic libraries, through storing genetic samples as tissue cultures, seeds, seedling and clonal plantations, to conservation of dynamic and evolving gene pools in natural forests. In tree improvement, where a common goal is to perpetuate a supply of high-quality wood fiber, genetic diversity is important in providing raw material for selection, in avoiding deleterious effects of inbreeding, and in guarding against threats of destruction by pests and pathogens. The choice of material for conservation is usually biased toward select phenotypes with desired commercial traits (Zobel 1977), or trees with broad adaptability (Ledig 1986). Germplasm for tree improvement is most often conserved in plantations and other kinds of gene banks. In the broader sense, when the goal is to perpetuate species per se, a combined approach of natural forest protection and gene banking is chosen (FAO 1975; Krugman 1984; Ledig 1986, 1988; Millar and Libby 1991). Selection of material in this case is intended to be random with respect to phenotype and genotype, and evolutionary resilience is sought. When the goal is to perpetuate forest trees as habitat for other organisms, the approach is to protect the entire natural community (Soule and Wilcox 1980; Harris 1984; OTA 1987). Trees are selected for their contribution to the desired ecosystem, and genetic aspects may be only one of many conservation criteria (e.g., demographic structure, population sizes, species mixes, etc.). It is important to realize that there are many methods of conservation, and they are not necessarily mutually exclusive. In practice, combined or integrated conservation methods will not only improve the success of individual programs, but will allow coordination among different types of conservation programs.

Efficient conservation activities for any of these goals benefit from the knowledge of genetic variation. This information guides the details of sampling and handling germplasm throughout the conservation process, including original collections and delimitations of land area, evaluation and monitoring, and distribution and use of the germplasm (Hawkes 1980). Fortunately for forest trees, a substantial literature is developing on basic genetic variability in species, and increasingly on the implications for conservation. Conservation of forest germplasm also stands to benefit from the lessons learned in agriculture, including the consequences of depleting native gene pools, and reducing variation within domesticated lines.

2 Use of Clonal Techniques for Germplasm Conservation

Whereas conservation strategies depend on knowledge of genetic variation for their conception, they depend on available technology for their implementation. New methods of clonal technology based on molecular and biochemical techniques are being developed, and their mastery broadens the range of conservation possibilities. Although clones are just starting to be used systematically, they have wide application in many conservation situations.

2.1 Advantages of Clones for Conservation

The advantages of using clones for genetic conservation relate to the ability to preserve and replicate exact genotypes without sexual recombination. The general consequence of this is to give control over genetic diversity, which is a critical tool for conservation (Libby 1983). Thus, although clonal forestry potentially threatens genetic diversity because monocultures of single genotypes can be propagated, the same opportunity to control diversity can be used advantageously to deploy appropriate levels of known diversity (Libby 1987; Leakey 1987).

Clonal propagation offers the only or the most important method of conservation in situations where seeds cannot be stored or where seeds physically are unavailable. The most obvious example is for species with recalcitrant seeds (seeds that have maximum storage life less than a few years; Bonner 1985, 1990). Many tropical tree species fall into this category, and also several important temperate genera, for example, *Quercus*, *Acer*, and *Aesculus* (Bonner 1990, and unpublished 1985).

In other cases, seeds may be produced under normal conditions, but may be unavailable at the time of collection. This may occur due to stressed conditions of the plants, including insect damage and drought, to demographic structure of the population (all plants juvenile), or to seed abortion following inbreeding. The clonal method allows these nonreproductive individuals to be propagated. Even when a stand is sexually mature and healthy, variation in seed production among trees within a stand and among years may create conditions such that collections limited to seeds would be disadvantageously small and/or nonrandom.

Clonal propagation is especially valued when the genetic integrity of seeds is suspect. Many opportunities for pollen contamination of natural stands, plantations, or gardens and arboreta exist – with possible sources of contaminating pollen being landscaping trees, forestry plantations containing trees of nonlocal origin, and genetically improved trees (Wilson 1990; Millar and Libby 1989a). In addition to genetic contamination, seeds may be genetically poor quality because of inbreeding depression. In these cases, clonal propagules avoid problems with seed, and, if the identity of standing trees is certified, assure that native germplasm is collected.

In addition to capturing otherwise unavailable genotypes, clonal propagation affords a security measure for conservation. In any germplasm collection or restoration program, replication is an important insurance against loss. In gene banking, replicates within a site and across sites are used whenever possible. Clonal propagation allows replications of exact genotypes, which is the best insurance.

An important phase of the conservation process is to evaluate the germplasm being protected – natural forest, plantations, tissue culture, or DNA. Evaluating the performance of accessions makes the germplasm collections valuable and useful. With clonal replication, genotypes can be characterized with greater statistical efficiency than through seedling progeny tests. Clones can be tested in many environments and their performance characterized by site (Libby 1983).

Clonal propagation is valuable for distribution and use of conserved germplasm. As information about performance accumulates on individual clones, they become more valuable, and distribution of exact genotypes is likely to be preferred (Libby 1983). For restoration, and for production forestry, certain tested clones can be prescribed from among the appropriate pool of native and local genotypes, to meet specific site conditions.

Obviously there are many ways in which clonal methods and methods based on sexual recombination can be used together for conservation. At some phase of a program, the safest route may be to use clones; at another phase, seeds may be cheaper and more reliable, and/or genetically more defensible. In other cases, the two methods may be used simultaneously.

3 Methods of Clonal Conservation

The methods of forest conservation are traditionally divided between in-situ and ex-situ approaches. I classify conservation methods by the packaging of the germplasm (e.g., tissue culture vs. whole tree; natural vs. artificial regeneration) and by the context of the germplasm collection (e.g., native forest, plantation, laboratory, or clonal archive). The diverse objectives for conservation, from preserving commercially valuable germplasm to conserving evolutionarily resilient populations, are met through combinations of these methods. Clones have application at every level, in circumventing biological barriers, and in providing alternatives and complements to methods with sexual reproduction.

The methods discussed below begin with tree germplasm conserved as wild trees in natural communities, and proceed through methods that conserve more reduced elements of germplasm, ending with DNA. In the former, the whole species plus associated forest ecosystems are conserved; in the latter, only fragments of germplasm are preserved in a highly artificial environment. Each method has advantages for certain goals, and strong programs are built by the integrated use of several methods.

3.1 Native and Near-Native Forests

Much germplasm-conservation effort calls for designation of in-situ forest reserves, where native trees are allowed to grow and evolve under natural conditions. The degree of management can vary from noninterventive (laissez-faire) to intensive. In strict reserves such as Wilderness Areas or Natural Areas, management is usually limited to suppressing catastrophic fires and prescribing manageable fires to reduce fuels and maintain target species. The opportunities for clonal manipulation are few, except in clonally reproducing species such as coast redwood (*Sequoia sempervirens*) or quaking aspen (*Populus tremuloides*), which are stimulated to sprout and coppice following fire. In these cases, a cleansing fire that stimulates clonal reproduction of existing trees will regenerate native germplasm on the site. This is preferred over seed regeneration if the pollen source is suspect, as in cases where trees of unknown or nonlocal source are planted nearby.

Gymnocladus dioica (Leguminosae) is a rare and endangered tree of north-central North America whose survival in natural stands has depended on its ability to reproduce vegetatively (Ambrose and Carey 1987). Although sexual reproduction occurs in populations in the United States, where within-stand levels of variation are high, most populations at the northern limits of its distribution in Canada are entirely clonal and genetically monomorphic within stands. Given the dioecious nature of the species, and the intensive land-clearing and agricultural development in its northern habitat, it is doubtful whether *G. dioica* would have survived without its ability to expand clonally. Despite the low levels of regional genetic variability, many of the Canadian populations are continuing to expand clonally, and appear vigorous (Ambrose and Carey 1987).

In some cases, reserves may be more intensively used and managed. Gene Resource Management Units (GRMUs, also called Genetic Conservation Areas) are a much-discussed but as yet little implemented form of land conservation (NCGR 1982; Krugman 1984) that allow manipulative use of the forest. The goal of GRMUs is to retain native tree species in populations with natural gene frequencies (Ledig 1988; Millar and Libby 1991). As long as these goals are likely to be achieved, activities such as timber harvest and artificial regeneration are allowed. Since these uses must not interfere with retention of native germplasm on the site, all planting must be done with the objective of approximating the natural genetic architecture. Clones allow the propagation of old-growth and other valued genotypes that would be lost following sexual recombination. Clones are especially important when the GRMU is surrounded by managed land with native species of non-native or unknown origin. Seed from native trees in the GRMU may be contaminated with pollen genes of this non-native origin. Clones allow regeneration of known native genotypes. However, if clones are used exclusively for long periods of time, the natural evolution of these populations will have been frozen.

In other situations, such as in state parks or national monuments, the objective may be to favor vegetation of a particular historic period. Clones afford a means for preserving old genotypes. In such cases, the freezing of evolution may be exactly what is desired.

3.2 Restored and Mitigated Forests

Of all fields in conservation biology, restoration science may emerge as the most important in the next half-century (Bradshaw and Chadwick 1980; Cairns 1980). The urgency to restore native ecosystems follows in the wake of rampant land development and conversion, deforestation, and pollution of the natural environment. Many natural resource agencies are restoring degraded sites, removing exotics, and reintroducing native species. Federal, state, and local mitigation laws require land developers to restore or transplant disturbed native communities to protected locations outside the development. Endangered species laws require that populations of listed species be recovered and that species be restored to self-sufficiency.

The need for restoration will be most urgent, however, if even a fraction of the projected consequences of global climate change occur (Peters and Lovejoy 1991; Peters 1990). Following such a change, vegetation patterns will be completely altered, and our knowledge of plant adaptation must by then be great if we are to reconstruct new and useful forest communities (Peters 1985).

The new field of restoration biology has grown primarily from ecology disciplines (Jordan et al. 1988). Restoration at the community and species level has been the main focus, with only minor attention to the genetic level (Millar and Libby 1989a). This situation is changing rapidly with the recognition that genetic source and genetic variability are important factors in plant survival and success on a site. With this awareness, agencies are incorporating genetic guidelines into their restoration policies (e.g., NPS 1988).

Clones have special value to restoration. They may be invaluable in rescuing endangered species, or any taxon for which census number is low. Since demographic stochasticity is a critical factor in the final stages of extinction, the most urgent goal in these circumstances is to increase plant numbers. Often this will be possible only with clonal expansion, since seed production may be absent or unreliable.

Clones are proving their effectiveness in the rescue efforts for several endangered species. For example, the Center for Plant Conservation, a national consortium of United States botanical gardens and arboreta, has used vegetative cuttings and tissue-culture methods to propagate wild material of the extremely endangered Catalina mahogany (*Cercocarpus traskiae*) (Reiseberg 1988). This species, endemic to Catalina Island in California, had declined in recent years to seven individuals, primarily because of habitat degradation and overgrazing from introduced goats, sheep, and feral pigs. Recent morphological and electrophoretic observations indicated that two of the seven remaining trees were hybrids with the common mahogany on the island, *Cercocarpus betuloides* ssp. *blanchae*. Cuttings from the pure Catalina mahogany are being propagated along with seedlings derived from wind-pollinated seeds from the island trees at Rancho Santa Ana Botanic Garden. Only the clones, however, are known uncontaminated genotypes. Both vegetative- and seedling-origin plants will be reintroduced into fenced habitat on the island.

Another species that may be aided by clonal propagation is giant sequoia (*Sequoiadendron giganteum*) (Fig. 1). The species per se is in no danger of extinction, as it is widely planted and protected by many in-situ and ex-situ collections. However, the northernmost native stand, the Placer Grove, is at risk. Disjunct from the rest of the distribution by 80 km, this tiny population of only six mature trees is genetically distinct and has the lowest within-stand genetic variation of all the groves (Fins and Libby 1982). Complicating conservation efforts is the presence of many young giant sequoias of uncertain origin that were planted in the grove by a local service group in 1951 (J. LaBoa, Tahoe National Forest, pers. commun. 1989). If pollen from any of these trees has contributed to the seeds of the mature trees, then the genetic integrity of naturally regenerated seedlings is compromised. Several conservation scenarios are being considered for the grove (W.J. Libby, pers. commun. 1989). The most prudent involves removing all planted sequoias and replanting with rooted cuttings of known native origin from existing clonal hedge banks. The hedge banks consist of plants from wind-pollinated families from five of the six native trees. Seeds that produced these hedges were collected before the contaminating sequoias were producing pollen.

Even widespread species may be threatened by extinction and require restoration. The American chestnut (*Castanea dentata*) has benefited from both natural cloning and ex-situ rooting of cuttings. Chestnut blight, caused by the foreign pathogen *Endothia parasitica*, all but eliminated the billions of chestnut trees that grew throughout the eastern United States. The species has survived only because of its ability to sprout from the base of killed stems. These sprouts grow for a few years, but typically become infected and die before they become

Fig. 1. Placer Grove of giant sequoia (*Sequoiadendron giganteum*), showing planted trees of uncertain origin in understory and mature native trees. This grove is the northernmost native population of giant sequoia, has only six surviving native trees, and differs from all other groves in several traits

sexually mature. Now, thanks to the chestnut's ability to sprout, there is hope for the species. A survey in 1984 (McKeen and Ambrose 1988) revealed individual chestnut trees that have regrown and remain blight-free. These clones are being tested for resistance, possibly maturation related, to *E. parasitica*, (J.D. Ambrose, University of Guelph Arboretum, pers. commun. 1988). If found to be resistant, such clones could be vegetatively propagated and reintroduced into chestnut habitat.

As in any clonal program, the danger of eroding genetic diversity is great in restoration, especially where disease resistance is involved. Often only a few individual survivors of a major epidemic remain in a species or population, or only a few resistant genotypes may be isolated. If these are clonally expanded and provide the basis for major restoration, then future problems may result from low diversity. Where possible, a phase of sexual reproduction should follow the initial clonal expansion of remnant or select wild genotypes. In some cases where the remnant individuals are related, inbreeding depression may reduce the vigor of sexually produced progeny. An innovative program for avoiding problems of inbreeding in captive propagation has been followed in the endangered Spekes gazelle (Templeton and Read 1983), where the captive population was forced through an early inbreeding bottleneck. This approach may also be useful in restoration of severely reduced tree populations or species.

The use of clonal propagation in mitigation allows the exact replication of genotypes in a new environment. If the site to be developed is well studied, the plants are mapped and cloned before development, and a new site is chosen that closely resembles the original, then it may be possible to reconstruct the original population successfully, notwithstanding nongenetic problems. Even in the case of many rare and endangered species or populations, where the original or mitigation site is atypical of native habitat, clones afford the opportunity to prescribe adapted genotypes for unusual sites (Libby 1983).

Another type of mitigation that may be done in non-native habitat would resemble a tree foster home. Species endangered in one part of the world would be established in similar environmental conditions elsewhere where protection was possible. The intent would be to maintain an option to reintroduce the genotypes back into the native habitat of the species once the threatening force was ameliorated. Clones could serve as the vehicle for moving genotypes and for regenerating trees in the foster forest. Clones would more accurately and easily maintain identities in the new site than would seeds unless controlled pollination were possible.

Forest decline in Europe provides a potential example. The nature of the threatening agent, air pollution, is difficult to escape in situ. Clones from the threatened species could be brought to North America and established in sites of similar physical conditions. Vegetative propagules from these trees could later be used for re-establishment in native European sites when air quality is improved.

Some tropical forests also are candidates for this type of rescue. Many areas are in imminent danger of exploitive deforestation. For many tropical species whose seeds do not store well, some form of clonal propagation may be the only means of propagating the species (Mathias 1988). Propagules from many of the species on a site would allow the structural re-creation of a tropical forest community valued for research and education as well as conservation. However, because many of the native elements would be missing from these foster forests, notably the pollinators, continued clonal regeneration might be necessary.

Clones will have at least two significant roles in restoration related to global climate change. The first should begin immediately and involves using clones to test genotypes for their breadth of climatic adaptability. Extensive evaluations will allow more effective restoration after climate change. Although there is general consensus that average temperatures will increase and precipitation decrease (NRC 1983), the direction of local climate change remains unpredictable (Hansen et al. 1988). This means that testing should evaluate adaptive niches in several climatic dimensions, although tolerance for warmier, drier conditions should be emphasized.

Adaptation can be evaluated in many ways, from routine progeny and provenance testing in tree-improvement programs to replicated plantations specifically for this purpose in gardens and arboreta. Both the Center for Plant Conservation in the United States (Thibodeau and Falk 1987; Falk 1990) and the Canadian Plant Conservation Program (Ambrose 1987a) are incorporating this element of off-site evaluation into their conservation programs.

Clones will also serve a primary function in actual restoration of areas where native vegetation could not tolerate climate change. Previously native species may be restored to the sites using different ecotypes adapted to the new conditions. Where clonal evaluation has indicated only marginal tolerance to the new climate, mixtures of clones from diverse but potentially adapted populations may be used to provide a pool of variation for evolution to develop adapted populations in the new sites.

3.3 Plantations

The goal in the examples described above is to maintain germplasm in natural or near-natural conditions. This includes native species, typical age-class mixes, and natural spacings and regeneration, with the result that the conserved forest serves as valuable habitat for plants and animals other than the target tree species. Furthermore, the species of this ecosystem continue to co-evolve in natural or semi-natural conditions. Plantations, however, have a much simpler structure with more controlled design and no opportunity to evolve. They consist of one or two species (common in temperate zone) to several species (common where agroforestry is practised), usually of the same age, and have regular spacing among trees. Competing species are undesired. Whenever germplasm collections are planted within the native range of the species, there is a risk of gene contamination from surrounding wild trees in the seeds of the planted trees. Using clonal replicates of the collection trees ensures that no contamination of the original germplasm occurs. The only way to ensure this with seedlings would be to make controlled crosses.

Plantations may be established specifically for germplasm conservation or they may primarily serve production forestry and only opportunistically or secondarily serve conservation. Plantations designed specifically for germplasm conservation of commercially valuable tree species usually complement ongoing domestication programs (Wood and Burley 1980). The degree of a program's allegiance to domestication can be measured by the nature of the conservation collections. Clonal archives that closely complement tree-breeding programs include clones with traits of immediate value to breeding. The diversity conserved in such collections may not represent allele frequencies in natural stands.

An example of clonal archives maintained to support economically valuable species is the National Clonal Germplasm Repository, part of the National Plant Germplasm System. It was established for the express purpose of maintaining valuable germplasm of agricultural crops (Council for Agricultural Science and Technology 1985; Parfitt 1988). The five existing repositories clonally propagate plants, and grow them in the field and in greenhouses. All accessions are stored on the USDA Genetic Resources Information Network (GRIN). Germplasm is obtained from existing plant-breeding programs and also from sponsored collection trips. For some species, a virus-testing program has been developed, and isozyme electrophoresis is used as an adjunct to morphological descriptors to characterize the accessions. Germplasm maintained in this program is distributed as scionwood.

Examples of clonal plantations specifically for germplasm conservation of forest trees include Monterey pine (*Pinus radiata*), giant sequoia, and coast redwood programs (Libby 1990; Fig. 2). In each of these cases, efforts were made to collect random samples of the population that represent the natural genetic structure of the species. The giant sequoia clones originated as wind-pollinated seeds from native stands whose seedlings were clonally expanded with rooted cuttings; the redwood clones were collected primarily as stem cuttings from seedlings in the wild. In the redwood collection, some wild sprouts were rooted and subsequently propagated through tissue culture to achieve some rejuvenation (or at least some reinvigoration – see Bonga and von Aderkas, Chap. 12, Vol. 1) and then as rooted cuttings. All species are maintained as plantations of hedged plants, from which cuttings may be taken. The redwood program includes 180 clones from 90 populations with clonal gene-conservation plantations at two sites in California and a third planned (J. Kuser, Rutgers University, pers. commun. 1988).

Species of *Populus* also have been the focus of specific conservation efforts. The Ontario Ministry of Natural Resources in Canada maintains clonal hedge archives of four commercially important native species (D. Rogers, OMNR, pers. commun. 1988). Sampling of the species was done in a grid pattern in the late 1970s. Some of the clones from the archive have been used in the breeding program.

Fig. 2. Genetic-conservation hedge orchard of giant sequoia, University of California Russell Reservation at Lafayette, California. Samples of nearly half of the 84 native populations are maintained here and at a US Forest Service clonally replicated hedge orchard at Chico, California

The Poplar Research Center of Belgium has an active program of preserving the native *P. nigra* gene pool. With known native populations rapidly diminishing because of land conversion, a collection was made in 1959 of remaining native trees in Belgium, and these trees are periodically intercrossed. Resulting seedlings are screened and evaluated, and select clones are exported to the operational program (Strobl 1987).

Many clonal tree-improvement and general forestry plantations serve opportunistically as germplasm repositories. The basic dilemma when serving both tree-improvement and genetic-conservation objectives is that the goal of increasing gain through breeding requires a narrowing of the genetic base, which counters conservation objectives (Barnes and Burley 1987). Furthermore, selections for tree improvement are nonrandom whereas collections for long-term genetic conservation are best made randomly. Many tree-improvement programs, however, spend considerable time and effort in expanding their genetic collections beyond what would be necessary for current breeding, often at the expense of immediate gain (e.g., Kitzmiller 1976, 1990). Plantations associated with tree-improvement programs, such as provenance or progeny tests, maintain broad genetic diversity. For agricultural species, the IBPGR specifies these as field gene-banks, and systematically inventories them for useful germplasm (Withers and Williams 1986). Identities and pedigrees of single trees are maintained in these plantations and performance is evaluated, both of which increase the value of the germplasm to conservation. The relatively short duration of some of these plantations decreases their value, but since many of the clones in a clonal improvement program will be retained and planted in other sites, they are not necessarily lost.

There are many examples of such improvement and production programs that have used clonal opportunities to provide back-up material for conservation. In Germany, the Institute of Forest Genetics and Tree Breeding is making a concerted effort to add a genetic conservation element to already existing clonal plantations of Norway spruce (*Picea abies*) (H. Muhs, Institute of Forest Genetics and Tree Breeding, pers. commun. 1988). The basis for the collections is an international provenance experiment begun in 1964, which contained 1100 sample provenances. The clones were replicated at three sites and most have been evaluated for growth and isozyme traits. These provenance plantations have been intentionally maintained for conservation by thinning to favor random diversity rather than favoring trees with commercially valued traits.

In Yugoslavia, the Poplar Research Institute maintains a clonal archive of poplars that has a more restricted genetic base (Herpka 1987). Selection for disease resistance and growth is initially made on progeny of *Populus nigra* crosses. The best clones are planted yearly in genetic-conservation stool beds, which are felled and allowed to resprout every second year. These archives provide material for further selection.

Other examples of programs that use clones in conservation include the work on *Pinus caribaea* by the Oxford Forestry Institute (OFI). In 1963, the OFI began extensive seed collections of *P. caribaea* for provenance trials (Barnes and Burley (1990). The seeds were widely distributed and the plantations were periodically evaluated. By the late 1970s, the program had progressed to a stage where a second round of seed collections was planned. In the intervening years, however, many of the original native stands either had suffered serious genetic depletion or had become inaccessible for security reasons. This catapulted the status of the original provenance plantations into being extremely valuable and in some cases irreplaceable germplasm collections. Subsequently, many of the trees in the provenance plantations were vegetatively propagated and distributed to regional centers throughout the tropics. Ex-situ conservation of this germplasm in subsequent breeding programs has been practised by applying theoretical concepts of multiple-population breeding (Namkoong et al. 1980) to practical plantation management (Barnes 1986).

In the past, botanical gardens and arboreta focused on display, education, and recreation as primary goals, and conservation was either ignored or accomplished as a consequence of these primary activities. More recently, many botanical gardens and arboreta are shifting emphasis such that conservation has become a primary goal (Simmons et al. 1976; Frankel and Soule 1981). In the past, garden plantings have included only a few individuals from one or two populations of each species. There is increasing recognition of the importance of appropriate genetic sampling for conservation collections in botanical gardens. The Center for Plant Conservation (CPC), with headquarters at the Arnold Arboretum near Boston, Massachusetts, is a consortium of 19 gardens and arboreta dedicated to genetic conservation of rare plants (Thibodeau and Falk 1987; Falk 1990). Gardens affiliated with the CPC are chosen for their ability to maintain and propagate conservation collections. Each garden is charged with protecting specific endangered taxa. Although the CPC gardens maintain ex-situ collections, mostly in the form of planted individuals, the focus of the program is on integrated strategies that transcend the dichotomy of in-situ versus ex-situ approaches (Falk 1987). Clonal propagation is commonly used for transferring germplasm from native sites to gardens, for expanding clones to be used in evaluation and dissemination, and for returning germplasm to the wild as restoration or mitigation plantings.

The CPC works with extremely endangered taxa, including many woody plants in addition to *Cercocarpus* discussed above. For example, the CPC is focusing on the Florida torreya (*Torreya taxifolia*), which has experienced a serious decline in recent years. These large trees once covered thousands of hectares in Florida and Georgia, but the species now stands on the brink of extinction. An interaction of environmental changes and a suite of fungal diseases threatens the species. The only remaining living plants arise as sprouts from the roots of top-killed adults but none lives to sexual maturity. Because disease is rampant in wild stands, habitat protection alone will not save the species. Together, the US Fish and Wildlife Service, The Nature Conservancy, and the CPC are mounting a multiphased project to save the species (McMahan 1989). A major aspect of this is collecting and propagating cuttings of wild plants from systematic population samples. These will serve as donors for subsequent clonal propagation and distribution to select botanical gardens and arboreta for research and conservation.

The major focus given by the CPC to the importance of genetic aspects in conservation has been instrumental in educating garden and arboreta staff about genetic conservation. In 1989, the CPC sponsored a conference on the Genetics and Population Biology of Rare Plants, which had the goal of summarizing current knowledge on the importance of genetic variation for ex-situ conservation (Mlot 1989; Holsinger and Falk 1991).

Other garden consortia and individual arboreta have similar goals. The Canadian Plant Conservation Program, based at the University of Guelph Arboretum, Ontario, also focuses on the importance of genetic variation in ex-situ collections and plantations (Ambrose 1987a). The Canadian program emphasizes regionally rare plants of Canada's Carolinian zone, where species of the eastern North America deciduous forest reach their northern distribution limit. Ex-situ conservation stands that have been established through vegetative propagation include such woody taxa as *Castanea dentata*, *Gymnocladus dioica*, and *Rosa setigera* (J.D. Ambrose, University of Guelph Arboretum, pers. commun. 1988).

Even gardens that have not identified conservation as a major focus often house valuable germplasm. Gardens have traditionally collected unusual variants and cultivars of species, many of which are propagated clonally. Together these can present a large opportunistic store of variation in a species. In Port-Orford cedar (*Chamaecyparis lawsoniana*), for instance, over 200 distinct cultivars have been described and are clonally propagated (Dallimore and Jackson 1966). Many derive from wild origins. Because they are widely planted and recognized for ornamental purposes, their performance in a wide range of environments could be assessed, and a profile of the adaptive breadths for each cultivar described.

Such a pool of germplasm, although not originally intended for conservation, may be what will save Port-Orford cedar in the wild. Port-Orford cedar is endangered throughout its range by a fatal root disease caused by the fungus *Phytophthora lateralis* (Zobel et al. 1985). Several programs for control and recovery are underway, including a genetic resistance study, and a silvicultural program. Although to date resistance has not been found in the wild species, the ornamental cultivars may offer a useful pool of genetic variation. Not only is it a highly variable pool, but propagules have been repeatedly subjected to nursery conditions where various species of *Phytophthora* occur in abundance. Thus, if there is resistance, some selection may already have occurred in cultivation.

3.4 Regenerable Tissues

The tissue most commonly used for regeneration is the embryo contained in the seed. Tissues that are clonally propagated include cells, organs, and tissue masses, rootable stem and leaf cuttings, root sections that produce shoots, grafts, apomictic seeds, and embryoids encapsulated in matrix. Although these tissues are the vehicles for multiplying plants in any clonal conservation program, they may also serve as primary units for germplasm conservation.

The role of tissue culture in germplasm conservation is best illustrated by the comprehensive program of the International Board for Plant Genetic Resources (IBPGR, Withers and Williams 1986) for agricultural species. The IBPGR program cycles newly acquired germplasm through phases of multiplication, distribution, storage, characterization, evaluation, and disease indexing. To coordinate research and conservation efforts on agricultural crops, IBPGR maintains a computerized international database related to all aspects of in-vitro conservation (Wheelans and Withers 1984).

Clonal techniques in the IBPGR program start in the field, by collecting tissue directly into culture media (Yidana et al. 1986). Germplasm is transported to the laboratory, multiplied in vitro, then distributed to participating institutions for storage. Tissues are stored either in traditional culture under slow-growth conditions, or by cryopreservation. Tissues for slow growth are usually cultured on solid media, to which protectant chemicals are added, and stored at low temperatures for several years (Henshaw 1987, and references therein).

Tissues cultured in slow growth are used for active research; the tissues are cloned and distributed to researchers and breeders. Long-term base collections not intended for frequent use are kept in cryopreservation at the temperature of liquid nitrogen (− 196 °C; Henshaw 1987). Before storage, clones are screened for specific disease-causing organisms, especially viruses (disease indexing), and infected cultures are cleaned up through serial propagation and thermotherapy (Walkey 1978, 1985). Since cell and tissue cultures are not easily identifiable by morphological descriptors, they are characterized by biochemical methods, most commonly electrophoresis. Some testing for disease resistance or nutrient tolerance can be done in culture. More extensive evaluations are made when the plants are grown out.

Concern for genetic instability (Scowcroft 1984) in cultures has motivated considerable research in the IBPGR program. The risk of instability resulting from the release of somaclonal variation in in-vitro systems ranges from minor to major, depending on the species and on the culture system used. There is evidence for some natural somatic instability in all living tissues (D'Amato 1978; Larkin and Scowcroft 1981;Roberts and Ellis 1984; Walbot and Callis 1985; Ahuja 1987a). In addition, in-vitro conditions themselves may be mutagenic to plant cells (D'Amato 1978).

In-vitro conservation programs for trees are not as extensive or as comprehensive as those for agricultural crops. Although there have been many advances in developing in-vitro techniques for clonally propagating tree tissues (Bonga and Durzan 1987a,b; Ahuja 1987b; Hasnain and Cheliak 1986; Hanover and Keathley 1988, and Chaps. 7–15, Vol. 1), there have been fewer examples of applications to long-term storage and conservation (Wilkins and Dodds 1983).

Experiments with traditional cold storage on *Populus tremula* clones indicate that branches with dormant buds can be stored and later cultured (Ahuja 1989). After a year in cold storage at 0 °C, − 5, − 8, − 18, and − 80 °C, only the buds stored at − 5, − 8, and − 80 °C showed good growth and differentiation in culture. Slow growth of cultures by storing cultures at low temperatures has successfully extended the storage of cultures from a few weeks to 1 year in several forest tree species (Table 1; IBPGR 1989). Research has also advanced on many fruit and nut species, where subculturing at 1-month intervals has prolonged storage indefinitely (IBPGR 1989).

Storing woody germplasm in liquid nitrogen is an exciting yet little researched or applied method (Sakai 1985; Ahuja 1989). For many species, it may provide safe storage far beyond typical durations under normal conditions (as long as two centuries, Bonner, unpubl. 1985). Cryopreservation is relatively simple and inexpensive, as long as a steady supply of liquid nitrogen is available

Table 1. Tree species that have been successfully stored and regrown under slow-growth tissue culture conditions. (IBPGR database 1989)

Species	Conditions (temp. reduction range, °C)	Storage length	Recovery rate (%)	Database reference
Betula pendula	Solid medium 25 – 4	16 weeks	100	Meier-Dinkel
B. pubescens	Solid medium 25 – 4	24 weeks	100	Meier-Dinkel
Larix decidua	Solid medium 22 – 4	2–3 months	Low	Bonga
Picea abies	Solid medium 21 – 4	24 weeks	30	Simola; Santanen
Pinus radiata	Solid medium 24 – 5	4.5 years	95	Aitken-Christie
Populus deltoides	Solid medium + activated charcoal 22 – 4	1 year	80	Overton; Sebastian; Vermeir; Evers
Populus tremula	Solid medium 22 – 5	1 year	[a]	Douglas
Pseudotsuga menziesii	Solid medium 25 – 6	4 weeks	[a]	Vermeir; Evers

[a]Not available.

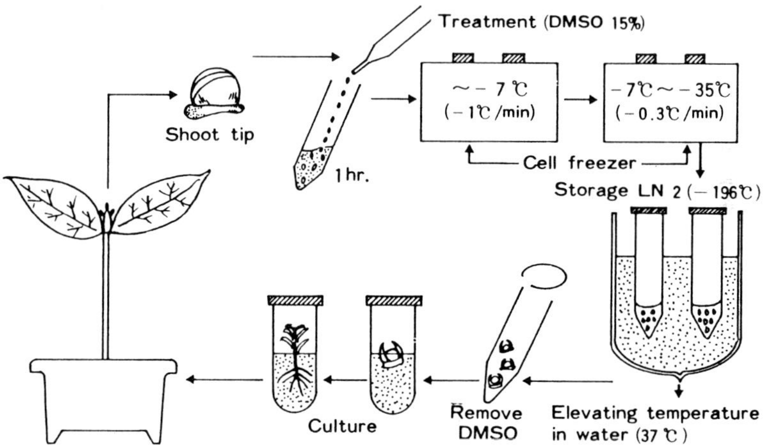

Fig. 3. Steps in cryopreservation of plant tissue, including preparation of tissues, slow freezing, storage in liquid nitrogen, thawing, and plantlet regrowth. (Reprinted with permission from the Institute of Forest Genetics, Republic of Korea)

(Fig. 3). Using appropriate cryoprotective compounds and rates of cooling and warming is critical to the survival of frozen cells (Finkle et al. 1985). Little is known about maximum possible lengths of storage, or the genetic damage that may accumulate in DNA during storage. Encapsulating excised embryos with artificial coatings may protect naked tissue against damage during freezing (Bonner, unpubl. 1985).

Considerable progress has been made on cryopreservation of agricultural plant tissues (Sakai 1985; Kartha 1985). IBPGR (1989) lists successful cryopreservation (i.e., living tissue recovered from exposure to ultra-low temperatures) for 60 agricultural species. Tree species included are angiosperms such as *Malus*, *Morus*, and *Prunus*, *Phoenix dactylifera*, *Ulmus*, *Populus tremula*, *Fraxinus*, and *Platanus* (Table 2). Much less work has been done on cryopreserving conifer tissues (Sakai 1985; Aitken-Christie and Singh 1987; Table 2). A method using cryoprotectants, prescribed freezing and warming rates, and specific container materials (Finkle et al. 1985) was successfully employed for *Picea abies* and *Pinus taeda* (Gupta et al. 1987). Research on *Picea mariana* and *Pinus banksiana* is underway (D. Rogers, Ontario Ministry of Natural Resources, pers. commun. 1988). Seeds of *Larix decidua*, *Pinus sylvestris*, and *Abies alba* have been successfully recovered from ultra-low temperatures (Ahuja 1989). Seed germination of seeds stored in cryopreservation conditions was over 90% of the control

Table 2. Tree species that have been successfully stored by cryopreservation (temperature of liquid nitrogen, −196 °C)

Species	Tissue	Storage length	Recovery rate (%)	Reference
1. Gymnosperms				
Abies alba	Seeds	6 days	100	Ahuja (1989)
Larix decidua	Seeds	6 days	100	Ahuja (1989)
Picea abies	Seeds	6 days	100	Ahuja (1989)
Picea abies	Embryogenic cells	10 min	50	Gupta et al. (1987)
Pinus sylvestris	Seeds	6 days	100	Ahuja (1989)
Pinus taeda	Embryogenic cells	10 min	50	Gupta et al. (1987)
2. Angiosperms				
Fraxinus	Callus	[a]	17–38	Shim (1987)
Malus spp.	Meristems	15 min	70	IBPGR (1989)
Malus spp.	Meristems	11 months	86	Katano (1986)
Morus spp.	Winter buds	1 month	50	IBPGR (1989)
Platanus				Sugawara and Sakai (1974)
Populus hybrids	Seeds	6 days	100	Ahuja (1989)
Populus hybrids	Callus	[a]	17–38	Shim (1987)
Populus tremula	Twigs with buds	2 years	14	Ahuja (1989)
Populus tremula	Excised buds	2 years	100	Ahuja (1989)
Prunus spp.	Shoot tips	24 h	[a]	IBPGR (1989)
Ulmus americana	Callus	4 min	42	Ulrich et al. (1984)
Phoenix dactylifera	Embryogenic callus	4 min	100 (after subculture)	Ulrich et al. (1982)

[a]Not available.

group. In many cases, storage was tested for a short time (2 days to 1 month), but tissue that can withstand short periods of supercooling should theoretically be able to tolerate much longer times (B. Finkle, Agricultural Research Service, pers. commun. 1988). In most cases reported, recovery of tissues was at least 50% the rate of the recovery from unfrozen tissues in the control group.

3.5 DNA

As molecular technology increases, we may expect to increase the usefulness of cloning and storing germplasm as DNA, either as whole genomes or as particular known and valuable sequences (Peacock 1984; Riemenschneider et al. 1988). Whole genomes can be stored as genomic libraries, and particular sequences can then be cloned when the need arises. DNA sequences are the most elemental form of germplasm that can be successfully preserved and recovered. Their usefulness depends on the ability to be transferred into a system where the genes can be incorporated and expressed. Sequences could be stored in bacterial plasmids or bacteriophage genomes. The recombinant molecules can then be cloned to produce potentially limitless copies of the stored sequence. Despite technical obstacles, progress is being made on transfer systems in higher plants including forest trees.

A major advantage of cloning and storing DNA for conservation is that DNA sequences potentially are transferable between species, and even among species in different kingdoms. The technology thus extends the use of known and valuable genes far beyond the reach of normal breeding. Transformation has been achieved to date in several forest trees (Ahuja 1988), including *Populus* species (Kriebel 1987; Chun et al. 1988), *Abies nordmanniana* and *Picea abies* (Clapham and Ekberg 1988), *Larix* (Diner and Karnowsky 1986), *Pinus densiflora* and *Quercus acutissima* (Choi et al. 1988), and *Pinus taeda* (Sederoff et al. 1986, 1987), where *Agrobacterium tumefaciens* was used as a vector. The work in these systems has been experimental and has not yet been applied to conservation.

Another advantage of DNA technology for conservation is the inert nature of DNA. This affords the theoretical ability to recover DNA from dead or extinct organisms. Although this technique has been successfully applied only in a few cases, the persistence of DNA suggests that the technique is possible for wide application. The inert nature of DNA makes it easy to store for conservation. Cloned DNA may be stored for indefinite periods by freezing at $-20\,^{\circ}C$ in appropriate buffer conditions (Mantell et al. 1985). For long-term conservation, DNA will most likely remain stable if genomic libraries are stored in inert state. In practice, metabolism of host vectors is reduced by cryogenic storage or freeze-drying, thus achieving a near-inert condition of DNA.

Recombinant DNA can also be maintained by normal culture of the vector bacterium. Appropriate measures must be taken to prevent genetic modification or loss of the sequence from the bacterium due to rearrangement and deletion of DNA, or to mutation followed by selection or genetic drift in the bacterial population. Instability in DNA collections occurs most likely in conditions where the host organism is maintained in an active condition. Plasmids are commonly lost spontaneously from bacteria. If this occurs in a portion of the population, part of the target DNA may be lost, in a manner equivalent to genetic drift. If there is selection against bacterial cells containing recombinant plasmids, these cells will be eliminated faster. Another opportunity for genetic change through drift may occur during amplification, when the host organism is cloned to high number to obtain abundant DNA. DNA rearrangements during bacterial cloning and deletion of target DNA may be a serious source of genetic erosion in genomic collections. This is especially problematic in plant genomes, which contain large portions of repeat DNA. Some alleviation of this problem has been achieved by using selection systems (Loenen and Blattner 1983).

An international program called DNA BANK-NET, with headquarters at Baylor University, Texas, was initiated in 1988 as an international network of DNA-conservation banks. DNA BANK-NET is founded on the premise that DNA transfer between species will be the most common method of crop improvement in the next century and that DNA diversity will be sought. Each acquisition is documented by an herbarium voucher, and information on the source, collector, and date of acquisition is recorded in a computerized database. DNA transcription by polymerase chain reaction allows disbursement of DNA to private and public groups. In its first decade, DNA BANK-NET will focus on acquiring rare and endangered tropical species. Although agricultural crops are the most important focus of the network, forest-tree germplasm is also included.

In extreme cases where a species is reduced to a few individuals or a species is threatened by devastating disease, drastic measures to induce genetic variation may be taken in an attempt to preserve the species at all. Molecular biology offers tools for this extreme situation. Genetic variation may be stimulated by increasing the frequency of transposition of genetic elements (Saedler and Nevers 1985). Such mobile pieces of DNA, common in bacteria, were found in higher plants first in maize (McClintock 1956), but have been found in other angiosperms (Mantell et al. 1985). DNA with sequences and behavior similar to transposable elements has been discovered in conifers (C. Kinlaw, U.S. Forest Service, pers. commun. 1989). There is evidence that transposition may occur in response to environmental stimuli, such as virus infection (McClintock 1984). Variants resulting from transposition appear to be more stably inherited than variants from other mutagenic techniques such as irradiation (Herpka 1987). As these elements become better understood, they may prove useful in inducing variation in species that are low in variation or dangerously threatened (e.g., chestnut). Obviously, induction of variation would be done only in extreme situations, under controlled conditions, and would always be followed by monitoring, selection, and testing.

4 Integrated Strategies for Conservation

Security is crucial to conservation, and diversification of methods is one way to increase security. Combined conservation of specific germplasm, for example, in DNA banks, in tissue culture, and in plantations or natural stands allows germplasm to be stored and used efficiently. Germplasm may be partitioned into base collections where germplasm preservation is highest priority, shorter-term active collections that are accessed frequently, and living collections in natural habitats where evolution proceeds. The time horizon for conservation varies from short to very long. For base collections, conservation activities ideally should be planned for centuries. This is difficult to achieve in a social environment of diminishing grant periods, scarce budgets, and transient public and institutional priorities. Dedication of even the longest projects often extends only to decades. Conservation projects that have existed for many years may be completely lost by even a short hiatus in funding (Libby 1990). An integrated approach, with germplasm replicated in different types of collections, increases the likelihood that the species being conserved will survive the many kinds of assaults that arise over the years.

Cloning technology offers the hope for rescuing neglected germplasm collections once support is secured. If collections are reduced in number or health, if seed production has been compromised by gene contamination, or if plants are still juvenile, then cloning will help to expand and replicate the bottlenecked germplasm. Although this does not increase the diversity of a compromised collection, it allows whatever is remaining to be secured and prepared for later augmentation of diversity through crossing, transformation, or mutagenesis.

References

Ahuja MR (1987a) Somaclonal variation. In: Bonga JM, Durzan DJ (eds) Cell and tissue culture in forestry, vol. 1 General principles and biotechnology. Nijhoff, Dordrecht, pp 272–285

Ahuja MR (1987b) The importance of regenerating systems in forest tree biotechnology. Ann For 13 (1-2): 13–23

Ahuja MR (1988) Gene transfer in forest trees. In: Hanover J, Keathley D (eds) Genetic manipulation of woody plants. Plenum, New York, pp 25–42

Ahuja MR (1989) Storage of forest tree germplasm at sub-zero temperature. In: Dhawan V. (ed) Application of plant biotechnology in forestry and horticulture. Plenum, New York, pp 215–228

Aitken-Christie J, Singh AP (1987) Cold storage of tissue cultures. In: Bonga JM, Durzan DJ (eds) Cell and tissue culture in forestry, vol. 2. Specific principles and methods: growth and development. Nijhoff, Dordrecht, pp 285–301

Ambrose JD (1987a) A rare plant programme for southern Ontario. In: Bramwell D, Hamann O, Heywood V, Synge H (eds) Botanic gardens and the world: conservation strategy. Academic Press, New York London, pp 341–343

Ambrose JD (1987b) What are we conserving? Public Garden 2(4): 9, 35–36

Ambrose JD, Carey CK (1987) *Gymnocladus dioica* in Canada: the biology of a rare tree at its northern limits of distribution. In: Abstr 14th Int. Bot Congr, Berlin pp 15–18

Barnes RD (1986) Multiple population tree breeding in Zimbabwe. In: Proc IUFRO Conf Joint Meet Working parties for breeding theory, progeny testing, and seed orchard, Williamsburg, Virginia 13–17 October 1986, pp 285–297

Barnes RD, Burley J (1987) Vegetative propagation for improved tropical trees. In: Abbott AT, Atkin RK (eds) Improving vegetatively propagated crops. Academic Press, New York London, pp 221–226 (416 pp)

Barnes RD, Burley J (1990) Tropical forest genetics at the Oxford Forestry Institute. In: Ledig FT, Millar CI, Riggs LA (eds) Conservation of diversity in forest ecosystems. For Ecol. Manag 35(1, 2): 159–170

Bonga JM, Durzan DJ (eds) (1987a) Cell and tissue culture in forestry, vol 1 General principles and biotechnology. Nijhoff, Dordrecht

Bonga JM, Durzan DJ (eds) 1987b) Cell and tissue culture in forestry, vol 2. Specific principles and methods: growth and development. Nijhoff, Dordrecht

Bonner FT (1990) Storage of seeds: potential and limitations for germplasm conservation. In: Ledig FT, Millar CI, Riggs LA (eds) Conservation of diversity in forest ecosystems. For Ecol Manag 35(1, 2):35–44

Bradshaw AD, Chadwick MJ (1980) The restoration of land: the ecology and reclamation of derelict and degraded land. Univ Cal Press, Berkeley, 317 pp

Cairns J (ed) (1980) The recovery process in damaged ecosystems. Science Publishers, Ann Arbor, Mich, 167 pp

Choi WY, Lee BS, Lee KJ, Chung HK, Lee SK, Shim SY (1988) Gene transfer into *Pinus densiflora* and *Quercus acutissima* by *Agrobacterium tumefaciens*. Res Rep Inst For Gen Korea 24:121–126

Chun YW, Klopfenstein NB, McNaab HS, Hall RB (1988) Transformation of *Populus* species by an *Agrobacterium* binary vector system. J Korean For Soc 77:199–207

Clapham DH, Ekberg I (1988) Induction of tumors by various strains of *Agrobacterium tumefaciens* on *Abies nordmanniana* and *Picea abies*. Scand J For Res:425–437

Council for Agricultural Science and Technology (ed) (1985) Plant germplasm preservation and utilization. US Agric Rep 106:1–35

Dallimore W, Jackson A (1966) A handbook of Coniferae and Ginkgoales. Arnold, London, 729 pp

D'Amato F (1978) Chromosome number variation in cultural cells and regenerated plants. In: Thorpe TA (ed) Frontiers of plant Tissue Culture. Int Assoc Plant Tissue Cult, pp 287–296

Davis MB (1991) Changes in geographical range resulting from greenhouse warming effects on biodiversity in forests. In: Peters RL, Lovejoy TE (eds) Proc World Wildlife Fund's Conf Consequences of global warming for biological diversity. Yale Univ Press, New Haven CT, in press

Diner AM, Karnowsky DF (1986) Agrobacterium, potential for genetic engineering in *Larix*. Res Dev Conf, TAPPI, Technology Park, Atlanta, pp 93–94

Dourado AM, Roberts EH (1984a) Chromosome aberrations induced during storage in barley and pea seed. Ann Bot (London) 54:767–779

Dourado AM, Roberts EH (1984b) Phenotypic mutations induced during storage in barley and pea seeds. Ann Bot (London) 54:781–790

Falk D (1987) Integrated conservation strategies for endangered plants. Nat Areas J 7(3):118–123

Falk DA (1990) Endangered forest resources in the United States: integrated strategies for conservation of rare species and genetic diversity. In: Ledig FT, Millar CI, Riggs LA (eds) Conservation of diversity in forest ecosystems. For Ecol Manag 35(1, 2):91–108

FAO/UNEP (eds) (1975) The methodology of conservation of forest genetic resources FAO, Rome, 127 pp

Finkle BJ, Zavala ME, Ulrich JM (1985) Cryoprotectant compounds in the viable freezing of plant tissues. In: Kartha K (ed) Cryopreservation of plant cells and organs. CRC Press Boca Raton, pp 75–113

Fins L, Libby WJ (1982) Population variation in *Sequoiadendron*: seed & seedling studies, vegetative propagation, and isozyme variation. Silv Genet 31:101–148

Frankel OH, Hawkes JG (eds) (1975) Genetic resources for today and tomorrow. Univ Press, Cambridge, 492 pp

Frankel OH, Soule ME (1981) Conservation and evolution. Univ Press, Cambridge, 327 pp

Gupta PK, Durzan DJ, and Finkle BJ (1987) somatic polyembryogenesis in embryogenic cell masses of *Picea abies* (Norway spruce) and *Pinus taeda* (loblolly pine) after thawing from liquid nitrogen. Can J For Res 17:1130–1134

Hanover JW, Keathley DE (eds) (1988) Genetic manipulation of woody plants. Plenum, New York, 519 pp

Hansen J, Lacis A, Rind A, Russell G, Fung I, Lebeduff S (1987) Evidence for future warming: how large and when? In: Shands WE, Hoffman JS (eds) The greenhouse effect, climate change, and US For Conserv Found, Washington DC, pp 57–75

Hansen J, Fung I, Lacis A, Rind D, Lebedeff S, Ruedy R, Russell G (1988) Global climate change as forecast by the Goddard Institute for Space Studies three-dimensional model. J Geophys Res 93(D8):9341–9364

Harris LD (1984) The fragmented forest. Univ Press, Chicago, 211 pp

Hasnain S, Cheliak W (1986) Tissue culture in forestry: economic and genetic potential. For Chon 62:219–225

Hawkes JG (1980) Crop genetic resources field collector's manual. Int Board Plant Genet Resourc Eucarpia 37 p

Henshaw GG (1987) New techniques for germplasm storage. In: Abbott AJ, Atkin RK (eds) Improving vegetatively propagated plants. Academic Press, New York London pp 303–313 (416 pp)

Herpka I (1987) Improvement of Aigeiros poplars in Yugoslavia. Ann For 13(1/2):39–49

Holsinger D, Falk D (eds) (1991) Genetics and population biology of rare plants. Univ press Cambridge

IBPGR (ed) (1989) In-vitro conservation database. IBPGR, Univ Nottingham School Agric, Sulton Bonington, Loughborough, LE 12 5RD, UK

Jordan WR III, Peters RL, Allen EB (1988) Ecological restoration as a strategy for conserving biological diversity. Environ Manag 12(1):55–72

Kartha KK (1985) Meristem culture and germplasm preservation. In: Kartha KK (ed) Cryopreservation of plant cells and organs. CRC, Boca Raton, pp 115–134, 267 pp

Katano M (1986) Seasonal changes of freezing tolerance of apple shoot tips. Proc Fac Agric Kyushu Tokai Univ 5:1–5

Kitzmiller J (1976) Tree improvement master plan for the California region. USDA For Serv, San Francisco, 123 pp

Kitzmiller JH (1990) Managing genetic diversity in a tree improvement program. In: Ledig FT, Millar CI, Riggs LA (eds) Conservation of diversity in forest ecosystems. For Ecol Manag 35(1, 2): 131–150

Kriebel HB (1987) Applying molecular biology to forest tree improvement. Ann For 13(1/2):3–12

Krugman SL (1984) Policies, strategies, and means for genetic conservation in forestry. In: Yeatman CW, Aftron DK, Wilkes G (eds) Plant genetic resources. Am Assoc Adv Sci Selected Symp 87:71–78

Larkin PJ, Scowcróft WR (1981) Somaclonal variation—a novel source of variability from cell cultures for plant improvement. Theor Appl Genet 60:197–214

Leakey RRB (1987) Clonal forestry in the tropics. Commonw For Rev 66(1):61–75

Ledig FT (1986) Conservation strategies for forest gene resources. For Ecol Manag 14:77–90

Ledig FT (1988) The conservation of diversity in forest trees. BioScience 38(7):431–439

Libby WJ (1981) What is a safe number of clones per plantation? In: Heybrook HB, Stephen BR, von Weissenberg K (eds) Proc 3rd IUFRO Worksh Genetics of host-parasite interaction in forestry, Sept 1980, Wageningen, pp 14–21 Sept 1980

Libby WJ (1983) Potential of clonal forestry In: Zsuffa L, Raute RM, Yeatman CW (eds) Clonal forestry: its impact on tree improv and our future forests. Proc 19th Meet Can Tree Improv Assoc Montreal, Canada pp 1–11

Libby WJ (1987) Testing for clonal forestry. Ann For 13(1/2):61–76

Libby WJ (1990) Genetic conservation of radiata pine and coast redwood. In: Ledig FT, Millar CI, Riggs LA (eds) Conservation of diversity in forest ecosystems. For Ecol Manag 35(1, 2):109–120

Libby WJ, Stettler RF, Seitz FW (1969) Forest genetics and forest tree breeding. Annu Rev Gent 3:469–494

Loenen WAM, Blattner FR (1983) Lambda charon vectors (ck 32, 33, 34) adapted for DNA cloning in recombination-deficient hosts. Gene 26:171–179

Mantell SH, Matthews JA, McKee RA (1985) Principles of plant biotechnology. Blackwell, Oxford, 269 pp

Mathias PJ (1988) Micropropagation of the tropical hardwoods, *Khaya invorensis* and *Nauclea diderrichii*. PhD Diss, Univ Nottingham, 312 pp

McClintock B (1956) Controlling elements and the gene. Cold Spring Harbor Symp Quant Biol 21:197–216

McClintock B (1984) The significance of responses of the genome to challenge. Science 226:792–801

McKeen CD, Ambrose JD (1988) Native chestnut attempts a comeback. Wildflower, Spring:20–23

McMahan LR (1989) Conservationists join forces to save Florida Torreya. Center Plant Conserv Newslett 4(1):1–8

Millar CI (1987) The California forest germplasm conservation project: a case for genetic conservation of temperate tree species. Conserv Biol 1(3):191–193

Millar CI, Libby WJ (1989) Restoration: Disneyland on native ecosystem? Restor Manag Notes 7(1): 18–23
Millar CI, Libby WJ (1991) Strategies for conservation of clonal, ecotypic, and disjunct population diversity in widespread species In: Holtsinger K, Falk D (eds) Genetics and population biology of rare plants. Univ Press, Oxford in press
Mlot C (1989) Blueprint for conserving plant diversity. BioScience 39(6): 364–368
Namkoong G, Barnes RB and Burley J (1980) A philosophy of breeding strategy for tropical forest trees. Trop For Pap 16. Dep For Commonw For Inst, Oxford, 67 pp
NAS – National Academy of Science USA (ed) (1972) Genetic vulnerability of major crops. Rep Natl Acad Sci USA Washington DC, 307 pp
NCGR – National Council on Gene Resources (ed) (1982) Douglas-fir genetic resources: an assessment and plan for California. California gene resources program, 274 pp
NPS – National Park Service (ed) (1988) Draft management policies, chap 5. US Natl Resourc Manag Direct
NRC – National Research Council (ed) (1983) Changing climate. Natl Academy Press, Washington DC, 496 pp
OTA – Office of Technology Assessment (ed) US Congress (1987) Technologies to maintain biological diversity. OTA-F-330. US Gov Print Office, Washington DC, 334 pp
Parfitt D (1988) The national germplasm repository: Davis, California. Diversity (13): 9–10
Peacock WJ (1984) The impact of molecular biology on genetic resources. In: Holden JHW, Williams JT (eds) Crop genetic resources. conservation and evaluation. Allen & Unwin London, pp 268–276
Peters RL (1985) Global climate change: a challenge for restoration ecology. Restor Manag Notes 3(2): 62–67
Peters RL (1990) Effects of global warming on forests. In: Ledig FT, Millar CI, Riggs, LA (eds) Conservation of diversity in forest ecosystems. For Ecol Manag 35(1, 2): 13–34
Peters RL, Lovejoy TE (eds) (1991) Proc World Wildlife Found Conf Consequences of global warming for biological diversity. Yale Univ Press, New Haven CT, in press
Reiseberg LH (1988) saving California's rarest tree. Center Plant Conserv Newslett 3(1): 1–8
Riemenschneider DE, Haissig BE, Bingham ET (1988) Integrating biotechnology into woody plant breeding programs. pp 433–450 In: Hanover J, Keathley D (eds) Genetic manipulation of woody plants. Plenum, New York, pp 433–450 (519 pp)
Roberts EH, Ellis EH (1984) The implications of the deterioration of orthodox seeds during storage for genetic resources conservation. In: Holden JHW, Williams JT (eds) Crop genetic resources: conservation and evolution. Allen & Unwin, London, pp 18–37 (296 pp)
Saedler H, Nevers P (1985) Transposition in plants: a molecular model, EMBO 4(3): 589–599
Sakai A (1985) Cryopreservation of germplasm of woody plants. In: Bajaj YPS (ed) Biotechnology in agriculture and forestry, volume 1: Trees I. Springer, Berlin Heidelberg New York, pp 113–129 (515 pp)
Scowcroft WR (1984) Genetic variability in tissure culture: impact on germplasm conservation and utilization. IBPGR Tech, Rep, Rome, 41 pp
Sederoff R, Stomp A, Chilton S, Moore L (1986) Gene transfer into loblolly pine by *Agrobacterium tumefaciens*. Bio Technology 4: 647–649
Sederoff R, Stomp R, Gwynn B, Ford E, Loopstra C, Hodgskiss P, Chilton S (1987) Application of recombinant techniques to pines: a molecular approach to genetic engineering in forestry. In: Bonga JM, Durzan DJ (eds) Cell and tissue culture in forestry, vol 1. General principles and biotechnology. Nijhoff, Dordrecht, pp 314–329
Shim SY (1987) Forest tree improvement research in Korea. Inst For Genet, For Admin Korea, 43 pp
Simmons JB, Beyer RI, Branham PE, Lucas G, Parry VTH (eds) (1976) Conservation of threatened plants. Plenum, New York, 336 pp
Soule ME, Wilcox BA (eds) (1980) Conservation biology. An evolutionary ecological perspective. Sinauer, Sunderland MA 395 pp
Strobl S (1987) Poplar culture in Belgium. On Line 4: 12
Sugawara Y, Sakai A (1974) Survival of suspension cultural sycamore cells cooled to the temperature of liquid nitrogen. Plant Physiol 54: 722–724
Templeton AR, Read B (1983) Elimination of inbreeding depression in a captive herd of Spekes gazelle. In: Shonewald-Cox C, Chambers S, MacBryde B, Thomas W (eds) Genetics and

conservation: a reference for managing wild animal and plant populations. Menlo Park, CA Cummings, pp 241–262 722 pp

Thibodeau F, Falk DA (1987) Building a national ex-situ network: The U.S. Center for Plant Conservation. In: Bramwell D, Hamann O, Synge H, Heywood V. (eds) Botanic gardens and the world conservation strategy. Int Union Conserv Nat. Academic Press, New York London (367 pp) pp 285–294

Ulrich JM, Finkle BJ, Tisserat BH (1982) Effects of cryogenic treatment on plant production from frozen to unfrozen date palm callus. Plant Physiol 69:624–627

Ulrich JM, Mickler RA, Finkle BT, Karnofsky DF (1984) Survival and regeneration of American elm callus cultures after being frozen in liquid nitrogen. Can J For Res 14:740–753

Walbot V, Callis EA (1985) Rapid genomic change in higher plants. Annu Rev Plant Physiol 36:367–396

Walkey DGA (1978) In-Vitro methods for virus elimination. In: Thorpe TA (ed) Frontiers of plant tissue culture. Int Assoc Plant Tissue Cult, Calgory, pp 245–254

Walkey DGA (1985) Applied plant virology. John Wiley & Sons, New York, 329 pp

Wheelans SK, Withers LA (1984) The IBPGR international database on in vitro conservation. FAO/IBPGR Plant Genet Resourc Newlett 60:30–38

Wilkins CD, Dodds JH (1983) Tissue culture conservation of woody species. In: Dodds JH (ed) Tissue culture of trees. Croom Helm, London, pp 113–136

Wilson BC (1990) Gene pool reserves of Douglas-fir. In: Ledig FT, Millar CI, Riggs LA (eds) Conservation of diversity in forest ecosystems. For Ecol Manag 35(1, 2):121–130

Wilson EO (ed) (1988) Biodiversity. National Academy Press, Washington DC

Withers LA, Williams JT (1986) In-vitro conservation. IBPGR research highlight 1984–1985. IBPGR, Rome, Italy pp 1–22

Wood PJ, Burley J (1980) Ex-situ conservation stands. Silvacultura 8:158–160

Yidana JA, Withers LA, Lyndsey A, Ivins J (1986) Development of a simple method for collecting and propagating cocoa germplasm in vitro. Acta Hortic

Zobel B (1977) Gene conservation – as viewed by a tree breeder. For Ecol Manag 1:339–344

Zobel DB, Roth LF, Hawk G (1985) Ecology, pathology, and management of Port-Orford cedar (*Chamaecyparis lawsoniana*). USDA For Serv Gen Tech Note PNW-184, 161 pp

4 Clonal Forestry with Sugi (*Cryptomeria japonica*)

K. Ohba[1]

1 Introduction

Since ancient times, sugi (*Cryptomeria japonica* D. Don) has been one of the most important conifer species in Japan. In the People's Republic of China, descriptions of *Cryptomeria fortunei* Hooibrenk ex Otto et Dietr, are also found, however, it is not so popular in Chinese forestry (cf. Hashimoto 1983). In Taiwan, in 1896, the seeds of *C. japonica* were firstly introduced from Japan, and at present about 45 000 ha of plantations have been established in the mountainous region at an elevation of 800–2000 m above sea level (cf. Sakaguchi 1983). In Korea, as it is cold and dry in winter, the *C. japonica* plantations introduced in ca. 1915 are limited to the southern area of the country. There are now about 74 000 ha of plantation (cf. Hyun 1983). According to the description of the above authors, *C. japonica* seems to have very limited uses in Taiwan and Korea.

It is believed that in former times abundant sugi forests prevailed in Japan. As sugi has a straight bole with soft wood, which could be split using even the primitive implements of the time, it was used for home construction, wooden ships, construction in paddy rice fields, as well as for many articles of daily use. During the Muromachi and very early Edo eras (about 500 years ago), a new use of sugi wood gradually developed, namely for wooden barrels with bamboo hoops. This new transportation container resulted in a revolution in the brewing and marketing of Japanese rice wine, soybean sauce, soybean paste, and also in recycling nightsoil. Sugi wood was suitable for the production of house beams, rafters, and various kinds of boards because of its straight bole, ease of processing, and pleasant color and scent. Today, sugi wood is even more important. Moreover, sugi produces more total growth and bole growth than any other tree species in Japan, and is completely adapted to the natural conditions in this country. Sugi is also easily regenerated by planting seedlings or stecklings.

[1] Institute of Agriculture and Forestry, University of Tsukuba, Tennodai 1-1-1, Tsukuba, Ibaraki, 305, Japan

Clonal Forestry II, Conservation and Application
Ed. by M.R. Ahuja and W.J. Libby

2 Natural Distribution of Sugi and its Genetic Architecture

2.1 History and Natural Distribution

The present natural distribution of sugi is shown in Fig. 1. By pollen analysis, Tsukada (1980) determined that during the last glaciation (about 15 000 years ago, when the sea level was lowered by about 130 m), and the natural range of sugi was limited to refuge areas on Yaku Island, Wakasa Bay, and the Izu Peninsula, and possibly other areas as well.

Yaku Island is famous for its huge sugi trees with ages of more than 1000 years. Kawada (1940) speculated that the sugi in Japan might have originated on Yaku Island, spreading to the north by two routes, namely, along the Japan Sea and along the Pacific Ocean, Yaku Island is believed to have had some isolated natural sugi stands even in this last glaciation. After the withdrawal of the glaciers, sugi moved from the refuge areas to the north, primarily covering vast lowland areas of Honshu. The sugi forests have been greatly exploited during the past 1000 years, and at present it has a patch-like natural distribution, as shown in Fig. 1.

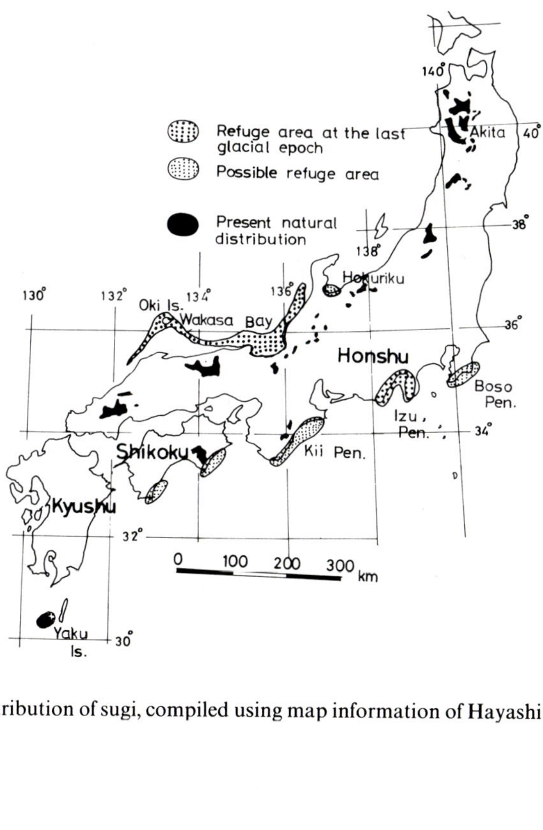

Fig. 1. Natural distribution of sugi, compiled using map information of Hayashi (1951) and Tsukada (1980)

2.2 Genetic Architecutre

Nakai (1941) described a variety of sugi (*C. japonica* D. Don var. *radicans* Nakai) with the Japanese name of Asio sugi. Murai (1947), after measuring needle length, needle curvature, and other features, first classified the sugi population into two groups as Ura sugi (from near the Japan Sea) and Omote sugi (from near the Pacific Ocean).

Distributional genetic differentiation was clarified by Sakai et al. (1978) by analyses of needle morphologies and peroxidase isoenzymes. They showed that one reproductive center of sugi is located in the areas of Yaku Island, Yanase (Shikoku), and Owase (Kii Penin.), and an additional center is in snowy Tohoku (Akita). Ogiyama and Yasue (1981) analyzed diterpenoids in the foliage of 57 natural sugi stands and classified them in the following four groups, namely, Akita, Hokuriku, Shikoku and Kii Penin., and Yaku Island. Yasue et al. (1987), using the above data made more detailed analyses and clarified two main lines of sugi, namely Ura sugi and Omote sugi. By a correlation coefficient of -0.77, the 57 natural sugi stands were separated into two major groups of Ura sugi (circle) and Omote sugi (square) respectively. Figure 2 shows the distribution of these two major groups on each side of Honshu Island, separated by mountains running through the island.

Ura sugi is distributed from the Yamagata Prefecture (northern limit) to the Tottori Prefecture (western limit), whereas Omote sugi has a wider distribution

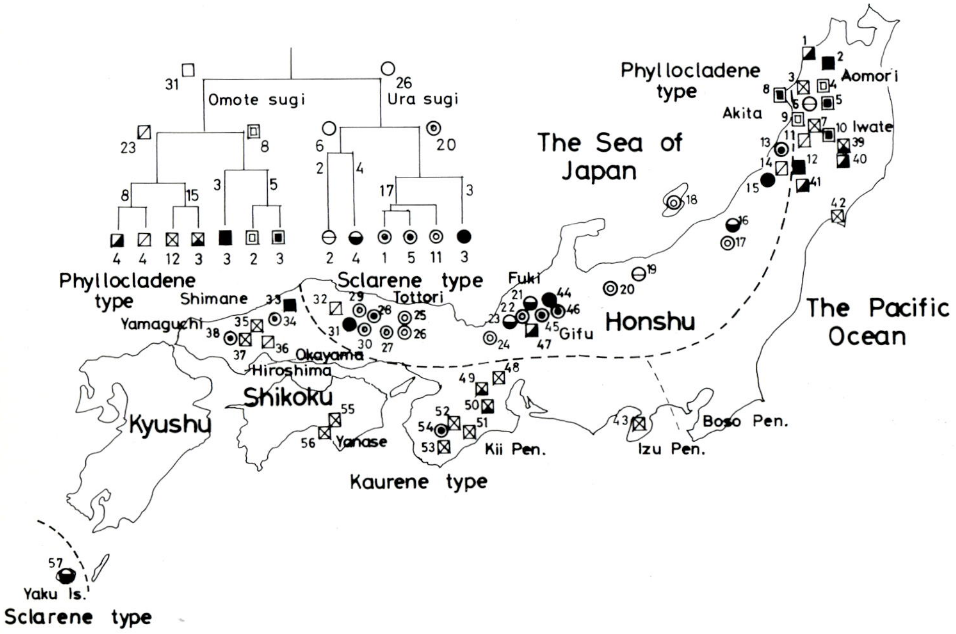

Fig. 2. Distribution map of 57 sampling sites of natural stands of *C. japonica* with classification symbols. *Circles* and *squares* with varied inner patterns indicate stands being classified by cluster analysis. *Circles* correspond to *Ura sugi* and *squares* show *Omote sugi* (Yasue et al. 1987)

from northern Honshu, from the Aomori and Akita Prefectures to the western Honshu Shimane and Yamaguchi Prefectures. Generally, it was believed that natural sugi stands in both Aomori and Akita Prefectures might belong to Ura sugi. However, it has been shown that they have an element of Omote sugi, shown by their diterpene hydrocarbon composition. The same was also true for sugi stands in southwestern Honshu.

3 A History of Sugi Cultivation and of Cultivars Propagated by Cuttings

It is believed that artificial planting of sugi began more than 500 years ago in the Kyoto districts and other southern areas of Japan. As planting materials, three kinds of propagation methods were used, seedlings, stecklings, and unrooted cuttings planted directly in the forest (hereafter: direct cuttings). Toda (1974) reported that, although it had been known for more than a dozen centuries that sugi and other conifers can easily be propagated by cuttings, forest plantations were not established by this method until the beginning of the 15th century. The oldest record is for the Kitayama forests near Kyoto city, where sugi stecklings have been planted successively since around 1400. In 1697, Miyazaki (Tokugawa 1947; Iura and Fukao 1979) published guidebooks for agriculture and forestry. These guide books included descriptions of how to raise sugi stecklings, giving practical details such as nursery-bed preparation, date of cutting collection, selection of cutting-donor trees, cutting preparation, setting the cuttings in the nursery bed, protection from frost, weeding, fertilization, and other useful information. Miyazaki also discussed the use of the unrooted cuttings planted directly in the forest. Tokugawa (1947) made an extensive review of the technologies of the Edo era on how to raise stecklings and how to plant unrooted cuttings directly in the forest. It was said that the use of direct cuttings was initiated around (1615–1623) in Hyuga (Miyazaki Pref.) and later was also started in Satsuma (Kagoshima Pref.) and Higo (Kumamoto Pref.) by order of the feudalistic lords. Tokugawa (1947) described site preparation for scion preparation, seasons of direct cuttings, insertion depth of cuttings, guiding rod for direct cuttings, etc. Rooting percentages of the direct cuttings in the forest were recorded as 50–87%, with a mean of 57%. There were also reports of poor results, such as 4%, these variable results probably depending on differences in technologies and in environmental conditions.

There are now more than 200 vegetative (clonal) cultivars counted in southwestern Japan. Here, cultivar means a group of trees propagated by cuttings with seemingly similar phenotype. As most of these cultivars were established centuries ago, this might, in some cases, be a single clone, and in most cases might be a clonal mixture with unknown numbers, but not so many. However, owing to the difficulties of making precise comparative scientific tests, only simple catalogs of their practical traits were sporadically made, including experiences and observation. Recently, extensive work on their productive traits was carried out by Miyajima et al. (1979) and Mashimo (1983).

In Fig. 3, a schematic distribution map of sugi shows both natural populations regenerated naturally, and old and new local cultivars propagated by stecklings. Clonal forestry has developed in areas with warmth index more than 100 °C, month and annual precipitation of more than 1800–2000 mm, and,

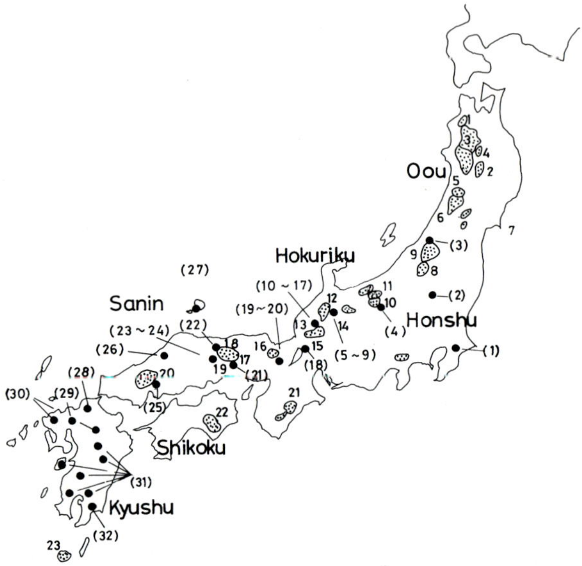

Fig. 3. Schematic distribution map of sugi for both local populations, mostly natural stands and local cultivars propagated by stecklings. (Miyajima 1983)

Local populations, mostly natural stands

1 Ajigasawa sugi
2 Ooshuku sugi
3 Akita sugi
4 Todo sugi
5 Chokaimura sugi
6 Yamanouchi sugi
7 Makinosaki sugi
8 Honna sugi
9 Mura sugi
10 Kuma sugi
11 Tateyama sugi
12 Hakusan sugi
13 Hanbara sugi
14 Itoshiro sugi
15 Mumai sugi
16 Ashiu sugi
17 Shiso sugi
18 Okinoyama sugi
19 Endo sugi
20 Hachiro sugi
21 Yoshino sugi
22 Yanase sugi
23 Yaku sugi

Cultivars propagated by stecklings

(1) Sanbu sugi
(2) Clones bred by Mr. M. Fukuda
(3) Mura sugi
(4) Kuma sugi
(5) Boka sugi
(6) Ryowa sugi
(7) Betsusho sugi
(8) Masuyama sugi
(9) Kawaidani sugi
(10) Kanemidani sugi
(11) Tateishi sugi
(12) Mizumi sugi
(13) Hekosan sugi
(14) Hatsuta sugi
(15) Ichinami sugi
(16) Itoshiro sugi
(17) Nyukawa sugi
(18) Tane sugi
(19) Shiro sugi
(20) Shibahara sugi
(21) Tomisu sugi
(22) Okinoyama sugi
(23) Endo sugi
(24) Tanami sugi
(25) Hachiro sugi
(26) Ichiki sugi
(27) Oki sugi
(28) Aya sugi
(29) Hon sugi
(30) Yabukuguri
(31) Measa
(32) Obi sugi

especially, with higher precipitation in spring to early summer. Old sugi plantation locations were dependent on river transportation of the wood and shifting cultivation, which reduced much of the cost of weeding.

Point number 21 (without parentheses) in Fig. 3, indicates the locations of the Yoshino sugi forestry where sugi plantations are established only by seeds collected from good quality aged trees more than about 70 years old that have been practically recognized to be high-quality trees by successive forest owners with their long expertise on the performance of the progenies. Dense planting of about 10000 seedlings per ha was common, and frequent thinning was made with a long rotation of ca. 80 years or more. Even at that time, there was a large market for thin logs in Kyoto and Osaka, and the mature trees were in great demand for wooden barrels, large and small. This is a good example of the success in genetic selection of superior trees in the areas where sugi regeneration with seedlings is common.

In the Meiji era (1868–), after the introduction of German forestry technologies, seedling planting became more popular and prevailed all over Japan, being strongly recommended by the forestry authority. Meanwhile, dysgenic selection was recorded in commercial seeds of sugi, especially in Yoshino sugi. As the Yoshino district became famous for superior sugi wood production, large amounts of seeds were ordered from many seedling growers, both private and official, throughout Japan. There were records of children collecting the sugi cones in the very young plantations for sale (cf. Toda 1969). Such dysgenic selection of seeds for plantations was quite possible for pines and hinoki, when the seeds were in great demand in the time from 1900 to 1930.

However, during a sudden outbreak of needle blight (*Cercospora sequoiae* Ellis et Everhart) in sugi around 1910, it was noted that stecklings were almost free from the disease, Chemical control was established for the disease, but vegetative propagation was re-evaluated, especially in those areas where they had used stecklings for regeneration; deployment of stecklings of local cultivars in respective areas became more popular again.

Unfortunately, similar experience in evaluation of individual seed trees of pines such as *Pinus densiflora* and *P. thunbergii* are lacking, because they regenerate well by natural seeding, and their wood is not so commonly used for house construction, which requires good wood quality. In hinoki, *Chamaecyparis obtusa*, the second most important tree species in Japan, natural regeneration was very difficult and hinoki has been regenerated by seedling planting. In comparison with sugi, less attention was paid to individual seed trees. However, there is only one instance in which clonal forestry was made by direct cuttings. They are now planted by stecklings called Nango-hi (Nango is a local area in Kumamoto Prefecture, hi means hinoki). This cultivar was first described by Sato and Miyajima (1956). The oldest existing plantation is about 120 years old, and the total plantation area of Nango hi is now about 8 ha (Shingai, pers. comm.). It is said that there are two strains of Nango-hi, with large or thin branches. It has a straight, full bole, and is resistant to resinosis and hypertrophy. The wood is flexible and pale pink and creamy white in color.

Numbers (19) and (20) in Fig. 3 indicate Kitayama forestry. This forestry practice originated with natural sugi populations that typically layered under the snow. The production of small polished logs of high quality became its main purpose, and in former days, the logs were grown successively in a group of sprouts of different ages, on the same stumps, called dai-sugi (sugi with successive selective cutting). After harvesting special stems that produced desirable boles in a clonal group on the same stump, they propagated their sprouts by cuttings and established several cultivars such as Shiro sugi, Honjiro, and others. The dai-sugi system is now hardly used (in the last 50–100 years), and stecklings of the selected clones are planted at a density of 5000 plants per ha, with rotation of 20–30 years and frequent pruning. In about 1893, the late Mr. Umeda (Iwamizu 1975) found a spontaneous mutant sugi that had a wavy bole surface. He propagated it by cuttings and kept the clone a secret. This type of polished wood was extremely valuable, with more than ten times the value of the common form. Now, a century later, more than ten popular wavy cultivars are grown, including Sango, Hirogawara, Ochiai, Ungai, and others named after their discoverers or the places where the mutants were found. Even now, this kind of mutant may be found at a frequency of roughly one out of several hundred thousand by observant private foresters across Japan (Iwamizu 1975). Meanwhile, Mr. Kato, who is a grower and processor of polished logs, makes extensive use of some fixed seed trees for his sugi plantations. Bole straightness and roundness with thin branches, and wood gloss are the main criteria of his selection (pers. commun.).

In Fig. 3, number (1) indicates Sanbu sugi forestry, where clonal forestry was established about 200 years ago. A clonal cultivar called Sanbu sugi was cultivated, which was thought to be a single clone under the nurse trees of Japanese black or red pines with a combination of crop cultures as in agroforestry. This cultivar has the traits of fast growth, straight bole, narrow crown, drought tolerance, and easy rooting, and was used for wooden ships, furniture, and many other kinds of articles for daily use. In some instances, a selective cutting system or hold-over system is carried out with 100–200-years-old trees of Sanbu sugi and hinoki.

Number (22) indicates Okinoyama sugi forestry. Here, clonal mixtures were propagated by cuttings from natural layers of old sugi in natural mixed forests. The artificial propagation of the clones is thought to have started around 1800. They collected the scions in the forests, but now some seedling growers have established cutting gardens with mixed clones, either many or a few.

Number (32) indicates Obi sugi forestry. The Ito family was invested with a fief in Obi in 1587. They selected suitable mountains, forests, and fields, and afforested them with sugi by means of direct cuttings. This practice allowed them to find a solution to overcome successive financial difficulties. This was said to be the beginning of Obi forestry. Later, this family, besides performing large-scale afforestation under direct management, rented forests to their dependents, and established an agroforestry system, whereby they shared the profit in the ratio of 50:50, which motivated the people to plant sugi. This ratio gradually changed to 33:67 in favor of private renters, then to 20:80, and most recently to 30:70 in the Meiji era, with the main share to the renters and the lesser share to the landowners. This became known as Obi forestry. In this region, the annual average temperature is 17 °C, and the annual amount of precipitation is between 2500 and 3000 mm. This climate is very suitable for the growth of sugi, and the registered increase in sugi growing here is one of the best in country. Namely, an average volume increment of sugi is about 12 $m^3 ha^{-1} yr^{-1}$ at 40-year rotation under average site conditions. Meanwhile, Obi sugi is superior with 17 $m^3 ha^{-1} yr^{-1}$. The Obi sugi wood was formerly mostly used for wooden ship building and was classified broadly into two lines, aka (red) and kuro (black), according to the heartwood color. As shown in Table 4a and b, the wood quality of Obi sugi is variable, according to the cultivars. At present, 15 cultivars are used. One of the requirements of Obi sugi cultivars is that they increase in volume rapidly at a uniform rate. In the Obi forestry system, vigorous growth was achieved by planting stecklings at 1000–1500 per ha. The use of sugi wood has now shifted more to house construction, and cultivation of Aka is now selectively used with its good wood quality and higher site adaptability.

4 Present Status of Sugi Cultivation with Special Reference to Cultivars

4.1 History and Composition of Clonal Cultivars

In Table 1 the current status of Japanese forests is shown and in Table 2 the area of plantations of sugi is presented according to age. Sugi is the most widely cultivated tree species in Japan. Namely, 4.51 million ha (44.2%) of the total 10.2 million ha planted is sugi, the second species being hinoki (*Chamaecyparis*

Table 1. Present status of forest resources in Japan. (Forestry Agency of Japan 1987)

	Area (1000 ha)	Growing stock (million m^3)		
		Conifer	Broadleaved trees	Total
Total	25 255	1785	1076	2861
Plantations	10 219 (40.5%)	1341	19	1360 (47.5%)
Natural forest	13 666 (54.1%)	444	1056	1500 (52.4%)
Others[a]	1 370 (5.4%)	0.3	1.3	1.6 (0.1%)

[a]Others: barren land, wet land, forest road, nursery, and others.

Table 2. Areas of sugi plantations by age class. (Forestry Agency of Japan 1987) (Unit: 100 ha)

	Total	Age classes (year)												
		1–5	6–10	11–15	16–20	21–25	26–30	31–35	36–40	41–45	46–50	51–55	56–60	60 <
Total	45 091	2131	3332	4651	7004	7865	7477	4970	1949	1359	1109	905	693	1646
National forest	8034	556	770	1080	1525	1181	1059	751	210	118	169	189	130	296
Private forest	37 057	1575	2562	3571	5479	6684	6418	4219	1739	1241	940	716	563	1350

obtusa (Sieb. et Zucc.) Endl.) with 2.34 million ha. As illustrated in Table 2, most of the sugi plantations are still young, less than 30 years old. They were extensively established after the World War II, replacing forests with broad-leaved tree species.

Forestry areas where sugi is grown naturally or extensively cultivated are shown in Fig. 3. The objectives of sugi wood production, length of rotation, planting density, and tending techniques are somewhat different in each sugi forestry area. To better illustrate the concept of sugi forestry, an example of sugi cultivation is shown in Fig. 4 (Funayama 1983).

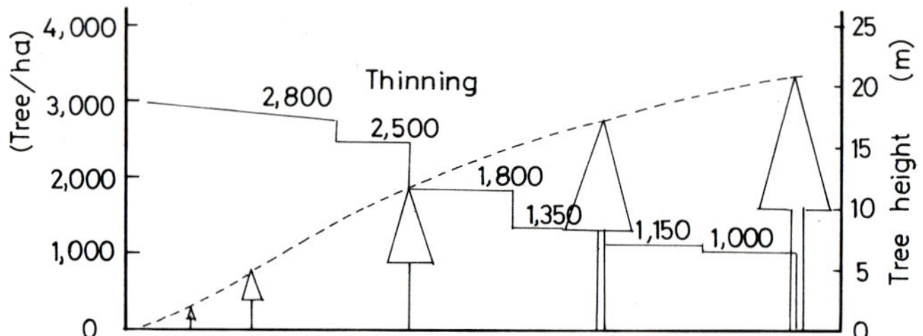

Stand age (yrs)	5	10	17	23	30	39	47	55
Mean height (m)	1.7	5.1	9.0	12.0	15.2	17.7	19.6	21.3
Mean DBH (cm)		6.8	11.1	15.1	19.5	23.0	26.2	28.0
Bole volume (m^3/ha)			140	240	360	450	520	580
Thinned volume (m^3/ha)			10	30	40	40	30	

Fig. 4. An example of sugi forest management. (Funayama 1983)

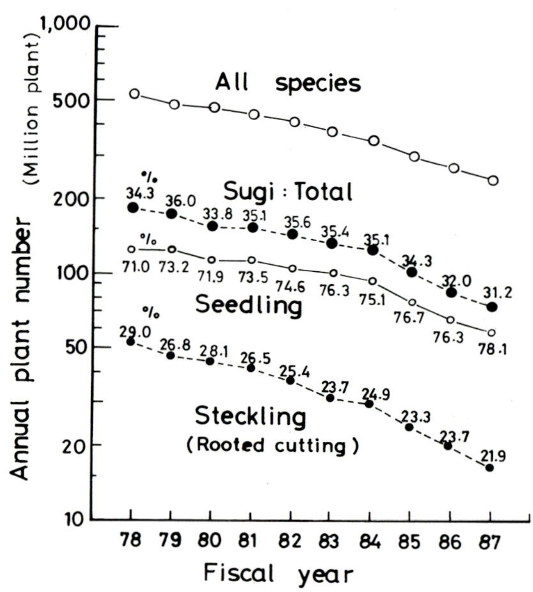

Fig. 5. Commercial production of sugi and other plants for artificial regeneration during the past 10 years. (Forestry Agency, Japan, unpubl. data, arranged by the author)

The commercial production of sugi plants for artificial regeneration over the decade 1978–1987 is presented in Fig. 5 (production in the National Forest is not included). As a result of severe economic conditions in Japanese forestry, the regeneration area decreased greatly during these decades, and nursery production of both seedlings and stecklings followed this decline.

Sugi represents about 35% of the total number of nursery plants produced annually. Stecklings accounted for about 25% of the total sugi stocks during this period. The mean production ratio of the stecklings of sugi for commercial use (cf. Fig. 5) for each Prefecture is presented in Fig. 6. Production of the stecklings dominates in southwestern Japan, especially in Kyushu. This reflects the distribution of sugi cultivars as shown in Fig. 3. A possible cause of the

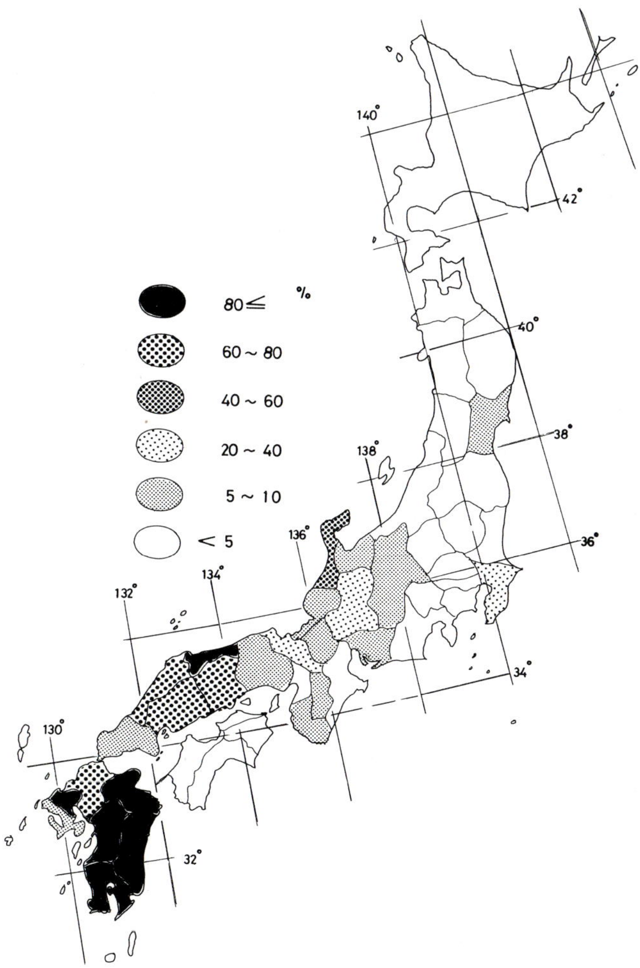

Fig. 6. Production ratio of commercial sugi stecklings to total sugi plants for regeneration during the past 10 years. Mean ratio is for each Prefecture. (Forestry Agency, Japan, unpubl. data, arranged by the author)

decreased trend from 29% stecklings of sugi in 1978 to 22% in 1987 may be a shift in planting more to the north by the policy of replacing broadleaved tree forests with sugi, where stecklings are not so commonly used.

4.2 Techniques of Clonal Propagation Now in Practice

Generally speaking, there are two modes for the production of stecklings. In Kyushu, 30–40-cm-long cuttings are rooted in open nurseries and, after 1 year, the stecklings are planted out in the forests. A success rate of 70% or more is expected in steckling production. In other areas, shorter cuttings of 10–20 cm are used in nursery or greenhouse culture, and after transplantation in the nursery for 1 to 2 years, they are planted out. The difference in the size of the cuttings is probably due to differences in climatic conditions between these areas; in Kyushu, it is warmer and therefore the growing season is longer than in the other areas. The cuttings were usually collected from so-called cutting donor forests—young forests of less than five to ten years old were planted with stecklings of respective cultivars. Less than five cuttings were taken from individual plants mostly once, and donor forests were changed each year. Because of the controlled use of superior cultivars and the expense of collecting cuttings from cutting donor forests, the cutting garden has become more important.

4.3 Clonal Deployment

Miyajima et al. (1979) reported that in the Kyushu area, about 500 years ago, old cultivars such as the Measa, Hon sugi, Aya sugi, and Yabukuguri strains, were originally established at shrines and temples as sacred groves or avenues, by unrooted cuttings planted directly in the forests or fields (direct cuttings). Therefore, people seem to have taken cuttings from these holy trees and distributed them in their local areas. Even now, there are huge sugi trees which were thought to be the cutting donor trees with ages of 200–500, or even more than 700 years. Miyajima et al. (1979) also reported examples of active cutting donor trees of the Measa strain of about 280 years old, having been established in 1688–1704 in Amakusa, Kumamoto Prefecture. The historical success of the clonal deployment of sugi in Kyusyu is primarily attributable to the instinctive success in the selection of sugi trees which had kept their good rooting ability even at such great ages by successive direct cuttings. The fact of this easy rooting of old cultivars resulted falsely in a general belief by Japanese foresters that all sugi could easily be regenerated by stecklings.

Since about 1956, there has been an active nationwide forest-tree breeding program based on about 3600 sugi plus tree selections. It has two major dimensions. One is the development of cutting gardens, with 522 ha established to date. The other is the development of classical clonal seed-orchards, with 570 ha now established. The cutting gardens are established with cuttings from the mature plus trees and serve mostly the clonal forestry areas in the south, while the seed-orchards are concentrated more in the north.

For the selection of sugi plus trees, it was decided that newly established cultivars (within the last 50 to 100 years) could not be included as plus trees, as these cultivars may consist of a single clone. There was no precise record of the propagation of the forests from which sugi plus trees were selected, except in the Kyushu district. In Kyushu, with the aim of clonal selection, many plus trees were selected from old cultivars which were thought to be clonal mixtures. Of 623 plus trees selected in the Kyushu district, 251 trees (40.3%) came from so-called steckling plantations, 264 trees (42.4%) from seedling plantations, and 108 trees (17.3%) are of unknown propagation. Clonal selection of the plus trees among and within cultivars has the advantage of well-known traits such as growth, bole straightness, adaptability to natural soil and climate conditions, wood quality, and others. They have been strongly recommended for planting.

Toda (1974) stated that at the beginning of the plus tree selection project with sugi, it was planned to supply commercial forestry with stecklings used as a mixture of many clones. However, vegetative propagation of sugi plus trees faced the difficulty, that cuttings from the majority of the mature selected plus trees showed very poor rooting ability. Therefore, the method of propagation was changed to the seed-orchard option, except in some districts such as Kyushu and Sanin, where regeneration using stecklings was still common from former times. In some areas, a vinyl house and mist spray are used to produce higher rooting for the propagation of these clones. Plant growth regulators, such as beta-indole butyric acid and its derivatives, are commonly used to improve rooting, especially for clones with intermediate rooting ability.

Another problem is the strong preference of private forest owners for homogeneous phenotypes as planting materials in the districts using clones. These growers do not like to regenerate their forests with clonal mixtures which express different characteristics.

About 30 years has passed since the beginning of the plus tree selection program; in the 1987 statistics of sugi planting, of 26489 ha, 8021 ha (30%) represents seeds with common seed forests, 2339 ha (9%) stecklings of so-called old cultivars, and 16129 ha (61%) with progenies of both seedlings and stecklings of plus trees. In sugi cutting production, of 18.9 million, 6.7 million (35%) is from common forests, 7.3 million (39%) is from old cultivars, and 4.9 million (26%) is from cutting gardens of plus tree clones.

Following the expansion of plantations, especially of sugi and hinoki, damage by climatic agents and pests increased noticeably. Climatic damage is mostly due to the expansion of sugi plantations to higher elevations by the expansion plan of sugi replacing natural broadleaved tree forests. In 1970, a breeding project for climatological damage was started, and more than 3000 sugi candidates with tolerance to frost, cold wind, and snow have been registered. Sugi breeding projects for resistance to sugi bark borer (*Semanotus japonicus* Lacordaire) and sugi bark midge (*Resseliella odai* Inoue) were also initiated in 1985. For sugi bark borer, sporadic damage has been noted locally since 1889 in Tokyo and Ibaraki. However, in the Kansai district, the damage of standing sugi bole has increased, and systematic surveys, mostly from the viewpoints of entomology and sugi wood damage, were started in 1955 (Research group of sugi bark borer, 1971). After that many local hot spots were found in Honshu Island. Sugi bark midge was first found in 1953 in southern

Kyushu and has spread all over Kyushu within the last three decades (Yoshida et al. 1980). The plus tree selection program is now facing a very difficult situation to cope with these pests. The clonal variation of old cultivars and plus tree clones were intensively studied for susceptibility to these insects, and only a few clones were found to be fairly resistant.

5 Characteristics of Clones Used for Afforestation

5.1 Clonal Composition of Accepted Cultivars

As shown in Fig. 3, many local cultivars are being regenerated by stecklings in practical forestry. The cultivars had been established by repeated vegetative propagations with or without selection, and it was not exactly known whether it was a single clone or mixture of the clones. It is more likely to be a mixture of clones with higher rooting ability.

Miyazaki and Sakai (1969) introduced the technique of isozyme analysis of sugi in Japan, Miyazaki and Miyajima (1981) made a thorough analysis of sugi cultivars in Kyushu on the basis of peroxidase isozyme patterns. They found that relatively few clones were mixed in each named cultivar. They examined 29 of the main native cultivars in Kyushu. Twenty-five of them showed distinct difference in isozyme patterns, indicating genetic diversity. Further analysis within each cultivar was done, taking samples from the huge sacred sugi trees from which these cultivars seemed to have descended, growing around shrines, temples, and other places scattered over Kyushu. The results were as follows: for Measa, eight different patterns in eight samples; Hon sugi, two different patterns in five samples; Aya sugi, three different patterns in five samples; and Yabukuguri, only one pattern in five samples. Judging from the peroxidase analysis, Yabukuguri, which has been widely used in Kyushu, seems to be a single clone.

A similar type of analysis was made by Taira (1979) in the Toyama district. Thirteen cultivars and one seedling population, with a total of 2736 plants, were analyzed for their peroxidase isozyme patterns. The number of peroxidase patterns varied within each cultivar, from at least two to little more than ten.

Okuizumi et al. (unpubl.) carried out isozyme analysis of sugi cultivars. Using six isozyme systems resolved into nine allozyme loci, one to five sample trees for each Obi sugi cultivar collected as genetic resources at Forestry and Forest Products Research Institute were analyzed. Six loci are identical genotypes and only in three loci of *Lap-1*, *6Pg-1*, and *Dia-3*, were genotypic variations with two to three types observed (see also Cheliak, Chap. 8, Vol. 1).

5.2 Clonal Traits and Relative Performance

Miyajima et al. (1979) described the phylogenetic analysis of sugi cultivars in Kyushu. Measa, Hon sugi, Aya sugi, and Yabukuguri have been proven to be old cultivars. Characteristics such as growth habit (fast or slow, early or late maturation), straightness of bole, forking, heartwood color (pink, red, black intermediate), flowering habit (mostly male, mostly female, and amount of flowers or fruits) have been closely investigated. In Fig. 7, site index curves of four representative cultivars are shown. Table 3 presents data from Mashimo (1983), which quantify the growth (site index) for various environmental factors including cultivars. From this table, it is possible to categorize the stand to be regenerated by the environmental factors, and make an estimate of the mean tree height at 40 years by summing up the scores corresponding to each category of six items. Soil type had the greatest effect on the estimated site index, with cultivars being second in importance.

The 12 cultivars were grouped as follows:

Growth	*Cultivar*
Very fast	Yabukuguri, Hon sugi, Urasebaru
Fast	Kuma sugi, Boka sugi
Intermediate	Ao sugi, Higo measa, Aya sugi, Obi sugi
Slow	Sanbu sugi, Okinoyama sugi
Very slow	Satsuma measa

In Table 4a and b, traits of 15 cultivars of Obi sugi in the Miyazaki Prefecture are presented (Miyazaki Forest Experiment Station, 1981). Each cultivar seemed

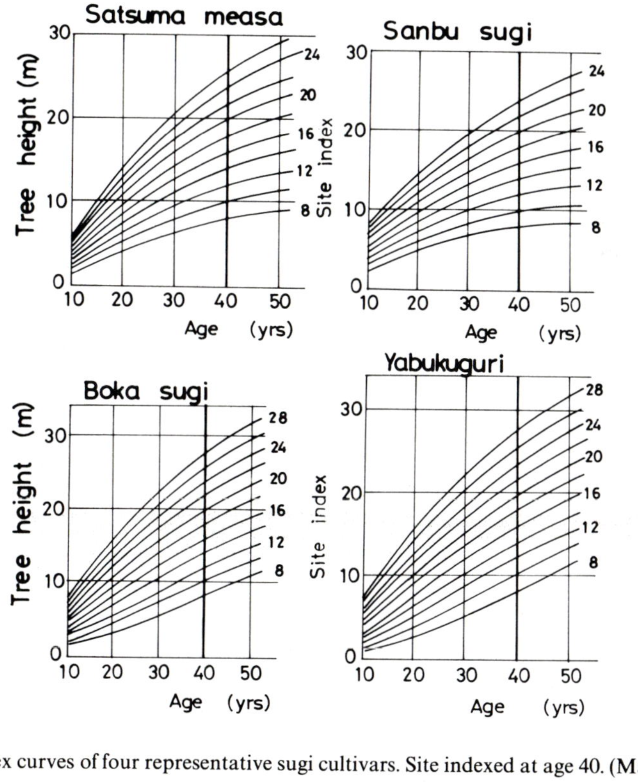

Fig. 7. Site index curves of four representative sugi cultivars. Site indexed at age 40. (Mashimo 1983)

Table 3. Quantification of growth (site index) of sugi cultivars for environmental factors

Factor Item	Category	No. of stands surveyed	Mean site index	Scores	Maximum difference of scores	Partial correlation coefficient
X_1, soil type	1 B_A, B_B, B_C	19	9.84	3.030		
	2 B_D (d)	43	14.26	7.567		
	3 B_D, deposit	21	17.19	9.451		
	4 B_D, creeping	68	20.15	11.893		
	5 B_D, colluvial soil	37	21.95	14.172	11.997	0.885
	6 B_E	41	23.17	15.027		
	7 B1, plate	24	13.54	5.589		
	8 B1, creeping	18	17.94	9.822		
	9 B1, colluvial soil	13	21.00	12.338		
X_2, exchangeable Ca	1 < 3 me/100 g	84	14.87	0.0		
	2 3–6 me/100 g	96	17.91	0.503	1.204	0.250
	3 > 6 me/100 g	104	21.60	1.204		
X_3, geology	1 Volcanic ash	80	15.94	0.0		
	2 Andesite	51	21.00	0.398		
	3 Granite	41	16.78	0.016		
	4 New sedimentary rock	43	20.63	1.484	2.284	0.340
	5 Old sedimentary rock	57	18.75	0.949		
	6 Serpentine rock, others	12	18.67	– 0.800		
X_4, warmth index	1 < 80 °C	31	16.19	0.0		
	2 80–95 °C	82	18.54	0.771	1.548	0.259
	3 95–110 °C	122	18.93	0.618		
	4 > 110 °C	49	18.00	1.584		

X_5, precipitation index	1 < 140	27	16.26	0.0		
	2 140–170	36	20.33	3.173	3.502	0.533
	3 170–200	138	18.85	3.252		
	4 > 200	83	17.37	3.502		
X_6, cultivar	1 Satsuma measa	22	14.77	0.0		
	2 Obi sugi	37	18.05	2.392		
	3 Sanbu sugi	21	17.00	1.455		
	4 Boka sugi	35	21.09	3.487		
	5 Kuma sugi	14	18.00	3.963	4.776	0.625
	6 Okinoyama sugi	17	15.71	1.364		
	7 Urasebaru	25	22.08	4.311		
	8 Aya sugi	38	17.26	3.234		
	9 Yabukuguri	25	19.46	4.776		
	10 Hon sugi	15	18.93	4.469		
	11 Higo measa	19	17.79	2.738		
	12 Ao sugi	16	18.13	3.227		

(Mashimo 1983)

The scores are used to estimate the growth index at the age of 40 years in upper-storied tree height in m.

The procedure for estimation is as follows: if a site is selected for regeneration with one of the cultivars, for example, Yabukuguri, each score corresponds to the category of the site in six items including cultivar (X_1–X_6) as shown with the underlined numbers, and is added to give the sum 25.66, which is an estimate of the site index of this site in (m) at 40 years old.

Table 4a. Traits of 15 cultivars of Obi sugi. (Miyazaki Forest Experimental Station 1981)

Cultivar (English name)	Color of heartwood	Shape of trunk	Bark	Shape of branch	Branch arrangement
1. Aka (Red)	Rose	Straight and perfectly round	Smooth	Slender and short	Rather upright
2. Arakawa (Coarse bark)	Rose	Straight and nearly round	Coarse and deeply cracked	Slender, twigs do not fall	With many straight, horizontal twigs
3. Haara (Coarse needle)	Rose	Not perfectly round and tapered much	A little coarse and deeply cracked	Ordinary	A lot of twigs obliquely upright
4. Grain	Rose	Not round and moderately tapered	Slightly cracked and wide piece of bark	Bow-shaped, with a lot of lateral twigs	Rather thick or dense
5. Mizorogi (Furrowed trunk)	Rose	Irregularly uneven	Somewhat intorted	Ordinary	Straight, horizontal
6. Hiki	Rose	With crooked base and no straight wood	Thin and colored red and white	Big	Straight, horizontal
7. Tosaaka (Tosa red)	Rose	Straight and perfectly round	Smooth and with some blue	Slender and short	With many twigs, moderately big
8. Edanaga (Long branch)	Rose	Not round and moderately tapered	Coarse and deeply cracked	Very long and big	Projecting joint and rather upright
9. Tanoaka (Tano red)	Rose	Straight and perfectly round	A little smooth	Slender and short	Rather upright
10. Hidarimaki (Left twisted bark)	Red with some black	Crooked or intorted	Coarse and spiral to left	Slender and short	With few twigs, coarse short and small
11. Genbe or ginyomu (Personal name)	When young, light black; when mature, pink	Straight and perfectly round	With medium thickness, deeply cracked	Ordinary	With many horizontal or downward twigs, shade tolerance
12. Tosaguro (Tosa black)	Black-brown	Straight and perfectly round, but tapered	Thick and coarse	Big and long with many lateral twigs	Ordinary
13. Chirimentosa (Wrinkled bark of Tosa)	Black-brown	Straight and perfectly round, but tapered	Coarse, but slightly cracked	Big and straight	Rather upright and dense
14. Kuro (Black)	Black	Straight and perfectly round	Smooth and slightly cracked	Slender and long	Grow densely and horizontally
15. Karatsuki	Black	Straight and perfectly round, but tapered	Coarse and deeply cracked	With many big and long twigs	Horizontal or upright

Table 4b.

Cultivar (English name)	Shape and color of leaf	Wood quality	Fruiting, female	Color of leaf in winter	Other traits
1. Aka (Red)	Thick and short	Good	Very few	Reddish brown	Grow vigorously after 20 years
2. Arakawa (Coarse bark)	Rather sparse, light dark green	Not oily and fragile	Very few	Reddish brown	Grow vigorously when young and mature
3. Haara (Coarse needle)	Thick	Rather light and soft	Very few	Reddish brown	Does not grow so well
4. Garin	Bright dark green and light	Light and strong	A great deal	Evergreen	Tolerates dry soil
5. Mizorogi	Small and soft	Very bad	Very few	Reddish brown	
6. Hiki (Furrowed trunk)	Long, soft and yellow	Beautiful, but very fragile	A great deal	Reddish brown	Twigs do not wither even if they are suppressed; shade tolerance
7. Tosaaka (Tosa red)	Long, soft and yellow	Good	Very few	Always dark green	
8. Edanaga (Long branch)	Slender and soft	Light, hard and knotty	Ordinary	Reddish brown	Sticky and does not break easily
9. Tanoaka (Tano red)	Thick and short	Good	Few	Golden	
10. Hidarimaki (Left twisted bark)	Long and light dark green	Light and watery	Very few	Evergreen	Decay of the heartwood may occure after ca. 50 years growth

Table 4b continued

Cultivar (English name)	Shape and color of leaf	Wood quality	Fruiting, female	Color of leaf in winter	Other traits
11. Genbe or ginyomu (Personal name)		Very resinous and suitable for ship-building (Benkozai)	From 8 or 10 years after planting	Reddish brown	
12. Tosaguro (Tosa black)	Rather long, soft and light dark green	Hard and lustrous, suitable for building timber	A great deal	Always dark green	
13. Chirimentosa (Wrinkled bark of Tosa)	Very long, soft, dark green and tufted	Heavy and rather soft	A great deal	Always dark green	
14. Kuro (Black)	Long, rather thick and light brown	Suitable for building timber	A great deal	Always dark green	Grow well up to the height of about 30 m
15. Karatsuki	A little long and soft	Suitable for building timber	Few	Always dark green	Grow very badly

Table 4a, b was compiled by Mr. Noriaki Hosoyamada who was the former deputy director of Miyazaki Forest Experiment Station. Around 1923, Obi District Forest Office, Forestry Agency was said to have a manuscript on Obi sugi, and in 1936, a guide book of Obi sugi including description of nine cultivars was published by the office. Then Ishizaki (1965) made more detailed measurements of needle, branch, bark, heartwood ratio, and others on the 15 cultivars given above. Mr. Hosoyamada made compiled tables of these data with his own surveys of both literature and field work. (These facts were examined by Mr. Takayoshi Sanui, who is a researcher of the Miyazaki Forest Experiment Station by direct inquiries to Mr. Hosoyamada and Dr. A. Ishizaki, pers. commun.).

to be a mixture of clones established by unrooted cuttings planted directly in the forest in connection with the type of agroforestry as described in Obi forestry. Around 1923, the Obi District Forest Office, Forestry Agency is said to have produced a manuscript on Obi sugi, and in 1936, the Obi District Forest Office published a guide booklet of Obi sugi based on this manuscript. In this booklet, they described the silvicultural traits of nine cultivars and the names of only five cultivars, as in Table 4a, b. At that time, it was supposed to be the large, hitherto, unwritten accumulation of the experience of peoples who had been concerned with growing and using of Obi sugi. Later, Ishizaki (1965) made more detailed measurements of needle, branch, bark, heartwood ratio, and others. However, Table 4 was compiled by Mr. Noriaki Hosoyamada, former deputy director of the Miyazaki Forest Experiment Station. He compiled these data from his own surveys of both of literature and field work (these facts were verified by Mr. Takayoshi Sanui, an entomologist of the station, by direct inquiries to Mr. Hosoyamada and Dr. Ishizaki; pers. commun. to Mr. Sanui). These data are distributed by the station and accepted as authentic.

6 Prospects for Clonal Forestry of Sugi in Japan

6.1 Breeding and Testing

In the early stages of plus tree selection and breeding, selection of undamaged trees was an important point. However, healthy individuals do not necessarily guarantee resistance of either sexual or clonal offspring to possible damaging factors such as pests, diseases, snow, and frost. Forestry is carried out in uncontrollable field conditions for many years. As artificial protection, including chemical sprays for pests and diseases and frost, is often prohibitively expensive, research on genetics and the practice of tree breeding should be initiated for such problems of practical forestry. The destructiveness of each factor is variable, depending on climatic, ecological, and other conditions. Damage may result from interaction between many factors, such as physical damage, environmental conditions, genetic architecture of the population including maturation effects, and silvicultural practices including planting density, mixing of tree species, pruning, thinning, and others. The necessity of countermeasures will be determined by the degree of damage that is tolerable in each managed sugi forest. Ohba (1979, 1983a, b) and Ohba and Furukoshi (1981) discussed the field of forest-tree breeding, and proposed to make maps and tables showing the isolines of the degrees of damage due to such destructive factors. They might be called maps of limiting factors for forestry management. A program of plus tree selection and breeding can then be constructed, incorporating the risk of damage to clones and plantations. Cowling et al. (1977) described a well-defined strategy for Southern pine forests in the USA (Ohba 1984a). If genetic improvement is adopted as a component countermeasure, we can use the maps for decision making in setting breeding objectives, in choosing breeding districts, in planting selection of breeding materials, and in advanced-generation and clonal testing.

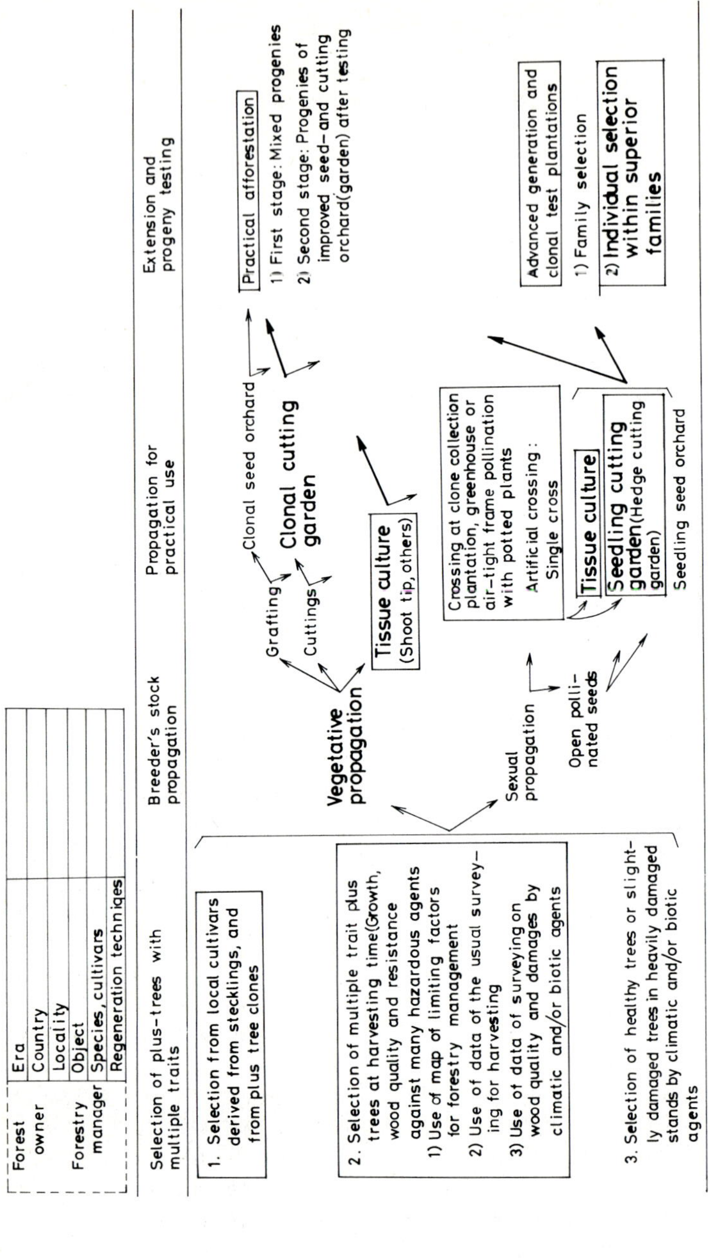

Fig. 8. A system of forest-tree breeding for resistance to local hazards by means of selection of multiple-trait plus trees. (Ohba 1984a)

In forestry management, it is hoped to provide a system of "Quality Control" (Ohba 1983a, b) similar to that common in industrial production processes. The system could function in each local area (breeding district) of each forestry management organization. It has become much more possible as systems of computers have come into common use. Management purposes, regeneration materials, tending techniques, and others, including tree-breeding activities, might be routinely assessed, and the results fed back to the respective operation in order to improve forestry management. Figure 8 illustrates how it is hoped to reduce the proportion of effort focused on classical progeny testing. To save time and to increase breeding effect per unit time, the selection of cultivars, and the selection of plus trees based on multiple traits at harvesting with the aid of the map information are of prime importance. Selection of very valuable trees and their mass clonal propagation might be practised under some of the optional schemes in Fig. 8.

6.2 Clonal Deployment

The most important fact that we should learn from preceding foresters in clonal sugi forestry areas, is their feed-back system of forestry management. There must be a clear relation that progenies raised by cuttings performed similarly to those of cutting donors only by experience or trial and error. Most of these foresters raised stecklings for their own uses, which was very different from the present situation using commercial stecklings. They were not only foresters but also farmers, fruit growers, horticulturists, even if the scale was small, and they had to be familiar with the productive traits of many crop plant species and so-called races or individuals. It must have taken long years to collect enough information and technologies from colleagues, family, or older villagers on clonal sugi cultivation. However, once this system had been established for a certain period, then they had the information and techniques to improve their management successively. The most important point of clonal forestry in former days and even nowadays is predictability of yield, quality of wood at expected ages, weeding, pruning, thinning, and others; and this predictability may become more clear for clonal deployment with a quality control system of intensive forestry management.

It is believed that for the genetic improvement of anemophilus trees, the use of vegetative propagation of trees with super phenotypes is one of the most efficient procedures. Burdon (1982) and Libby (1983) made comprehensive analyses of the advantages of clonal forestry in relation to selection, ease of management, and roguing cutting gardens, and to the option of grouping clones for particular adaptive areas or sites.

However, in many tree species, rooting ability decreases rapidly as the cutting donors mature by even 5 or 6 years, and subsequent growth is affected by cyclophysis and topophysis (see Greenwood and Hutchison, Chap. 3, Vol. 1). Meanwhile, it is well known that cuttings taken from young seedlings root easily, as shown in *Sequioa sempervirens* and *Pinus radiata* (Libby 1969), *Picea abies* (Kleinschmit and Schmidt 1978, Kleinschmit 1983), *Chamaecyparis obtusa* (Maebashi 1982). Phillon (1982) successfully mass-propagated *Picea glauca* by rooting cuttings taken from juvenile seedlings. In sugi, the so-called hedging technique was developed by Fujita (1931, 1933, cited in Toda 1969). Following rapid progress in the in vitro culture of several tree species, including sugi (Ishikawa, 1974), it may be possible to secure founder stocks from the best multi-trait plus trees at any state of maturation (see Bonga and von Aderkas, Chap. 12, Vol. 1). One of the most important fields of biotechnology in forestry might be explant propagation from superior ortets, whose state of maturation prevents vegetative propagation, by the simple rooting of cuttings. As shown in Fig. 8, an important strategy for future forest-tree breeding is to manage the risk of loss due to diseases and other damaging factors. As reported by Ohba (1978) and Libby (1982), clonal forestry with a single clone or a mixture of a small number of clones is not necessarily dangerous for any one damaging factor. One of the most important factors is the presence of a sufficient ratio of resistant clones with respect to all known and unknown factors that may damage them.

We are attempting to produce maps and tables indicating the limiting factors for forest management. A comprehensive approach for sugi cultivation is needed to make smaller breeding districts, a factor strongly favoring clonal forestry (see Ahuja and Libby, Chap. I.1, Vol. 2). Key points for success in the breeding phase might include efficient combinations of selection of healthy multi-trait plus trees, anticipating damaging agents, rapid mass-propagation by vegetative means (tissue culture and/or hedge cutting gardens of ramets and of tested full-sib families), and a follow-up system of the regenerated materials in practical forestry management with a quality control system providing feed-back to the breeders.

References

Burdon RD (1982) The role and optimal place of vegetative propagation in tree breeding strategies. In: Proc IUFRO Meet Working Parties Genetics about breeding strategies including multi-clonal varieties. Lower Saxony For Res Inst FRG, pp 66–83

Cowling EB, Dinus RJ, Schmidt RA (1977) Management of fusiform rust: A status report and recommendation for research. In: Dinus DJ, Schmidt RA (eds) Management of fusiform rust in southern pines. Symp Proc Gainesville, Univ Fla, pp 1–9

Forestry Agency of Japan (ed) (1987) Forest resources in Japan. Jpn For Tech Assoc, Tokyo, pp 1–77 (in Japanese)

Funayama Y (1983) Regional forestry and management. In: Sakaguchi K (ed) All of the sugi, *Cryptomeria japonica.* Natl For Extens Assoc, Tokyo, pp 530–562 (in Japanese)

Hashimoto Y (1983) Cryptomeria in Peoples's Republic of China. In: Sakaguchi K (ed) All of the sugi, *Cryptomeria japonica.* Natl For Extens Assoc, Tokyo, pp 594–599 (in Japanese)

Hayashi Y (1951) The natural distribution of important trees indigenous to Japan. Conifer report 1. Bull Jpn Gov Exp Stn, pp 48–240 (in Japanese)

Hyun K (1983) *Cryptomeria japonica* in Korea. In: Sakaguchi K (ed) All of the sugi, *Cryptomeria japonica.* Natl For Extens Assoc, Tokyo, pp 617–629 (in Japanese)

Ishikawa H (1974) In-vitro formation of adventitious buds and roots on the hypocotyl of *Cryptomeria japonica.* Bot Mag (Tokyo) 87:73–77

Ishizaki A (1965) Morphological, physiological and silvicultural characteristics of some important varieties of *Cryptomeria japonica* D. Don in Kyushu. Bull Jpn Gov For Exp Stn 180:1–303

Iura T, Fukao S (1979) Miyazaki A: several tree species, vol. 9. In: Yamada T, Iura S (eds) Nippon Nousho Zenshu 3:189–198 (in Japanese)

Iwamizu Y (1975) All of the polished log – new forestry goods production and commercial distribution of the polished log. Jpn For Assoc (Tokyo) pp 1–416 (in Japanese)

Kawada M (1940) Lecture on forest ecology (Ecological specificities of forests in Yaku Island and their management). Yoken-do (Tokyo) pp 1–528 (in Japanese)
Kleinschmit J (1983) Concept and experiences in clonal plantations of conifers. In: Zuffa L, Rauter RM, Yeatman CW (eds) Clonal forestry: its impacts on tree improvement and our future forests. Proc 19th Meet Can Tree Improv Assoc, pt 2, Toronto pp 26–56
Kleinschmit J, Schmidt J (1978) Experiences with *Picea abies* cutting propagation in Germany and problems connected with large scale application. Silv Genet 26:197–203
Libby WJ (1969) Some possibilities of the clone in forest genetic research. In: Bogart R (ed) Genetic lectures I. Oregon State Univ Press, pp 121–136
Libby WJ (1982) What is a safe number of clones per plantations? In: Heybroek HM, Stephan BR, von Weisenberg K (eds) Resistance to diseases and pests on forest trees. Proc 3rd Int Worksh Genetics of host-parasite interactions in forestry. PUDOC, Wageningen, pp 342–360
Libby WJ (1983) The clonal option. Norw For Res Inst, As-NLH Norway, pp 1–32
Maebashi Y (1982) A new technique of raising stecklings of hinoki, *Chamaecyparsis obtusa*. Tottori Pref Dep Agric For Fish Silvicult Div Jpn pp 1–32 (in Japanese)
Mashimo I (1983) Growth of plantations of sugi, *Cryptomeria japonica* and environments. In: Sakaguchi K (ed) All of the sugi, *Cryptomeria japonica*. Natl For Extens Assoc, Tokyo, pp 99–123 (in Japanese)
Miyajima H (1983) Cultivars. In: Sakaguchi K (ed) All of the sugi, *Cryptomeria japonica*. Natl For Extens Assoc, Tokyo, pp 126–140 (in Japanese)
Miyajima H, Nakao T, Nogami K (1979) Silvicultural characteristics of some important clonal cultivars of *Cryptomeria japonica* D. Don In: Miyajima (ed) Studies on characteristics of native clonal cultivars of *Cryptomeria japonica* in Kyushu. Kyushu Univ Press, Fukukoa, pp 1–66
Miyazaki Y, Miyajima H (1981) Identification of clone varieties in *Cryptomeria japonica* in Kyushu using peroxidase isozyme pattern. In: Callaham RZ (ed) Proc Div 2nd IUFRO World Congr Jpn IUFRO Comm, Ibaraki, pp 219–223
Miyazaki Y, Sakai K-I (1969) Use of zymography for identification of a clone of *Cryptomeria japonica* D. Don. J Jpn For Soc 51:235–239
Miyazaki Forest Experiment Station (ed) (1981) Obi forestry, point 20–2, Miyazaki prefecture. In: 17th IUFRO Congr, Kyoto, Jpn, Excursion 5: Forestry in the warm-temperate zone, Ibaraki, pp 1–9
Murai S (1947) Major forestry tree species in the Tohoku region and their varietal problems. Kokudo Saiken Zourin Gijutsu Kouenshu, Aomori-rin'yukai, pp 131–151 (in Japanese)
Nakai T (1941) Speciality of Asio experiment forest of Kyoto Imperial University. J Jpn Bot 17:273–283
Obi District Forest Office (ed) (1936) Obi sugi, pp 1–44 (in Japanese)
Ogiyama K, Yasue M (1981) Chemosystematic study of diterpenoid in foliage of *Cryptomeria japonica* D. Don. In: Callaham RZ (ed) Proc Div 2, 17th IUFRO World Congress, Jpn IUFRO Comm, Ibaraki, pp 225–227
Ohba K (1978) Presence percentage of plus-traits clones and their safety numbers for mixing. Jpn For Soc Kyushu Branch Res Pap 31:77–78 (in Japanese)
Ohba K (1979) Forest tree breeding in Japan, Jpn Agric Res Q 13:138–144
Ohba K (1983a) Risk evaluation and tree improvement as components of intensive forest management in Japan. For Ecol Manag 6:245–262
Ohba K (1983b) Cryptomeria in Japan. In: Zsuffa L, Rauter RM, Yeatman CW (eds) Clonal forestry: its impact on tree improvement and our future forests. Proc 19th Meet Can Tree Improv Assoc, Toronto, pt 2, pp 145–162
Ohba K (1984a) Genetics and breeding strategy of cryptomeria. In: Chopra VL, Joshi BC, Sharma RP, Bansal HC (eds) Genetics: new frontiers. Proc 15th Int Congr Genetics, New Delhi, vol 4: Applied genetics, pp 361–371
Ohba K (1984b) A new development of tree breeding by the use of techniques of clonal propagation. In: 5th Symp Basic breeding sciences, Gifu, Japan, pp 13–20 (in Japanese)
Ohba K, Furukoshi T (1981) Forest tree breeding and progeny testing in Japan. In: Prade J (ed) Proc Div 6, 17th IUFRO World Congr, Jpn IUFRO Comm, Ibarki, pp 175–178
Ohtomo E (1983) Management. In: Sakaguchi K (ed) All of the sugi, *Cryptomeria japonica*. Nat For Extcn Assoc, Tokyo, pp 486–529 (in Japanese)
Phillion BJ (1982) Large-scale production of black spruce cuttings for progeny tests. Paper presented in Intnal. Plant Prop Soc Meet, Grand Rapids, Mich, p 14
Research Group of Sugi Bark Borer (ed) (1971) Studies on the wood decay by sugi bark borer. Kansai Reg For Res Org, For Protect Div, pp 1–58 (in Japanese)

Sakaguchi K (1983) *Cryptomeria japonica* in Taiwan. In: Sakaguchi K (ed) All of the sugi, *Cryptomeria japonica*. Nat For Extens Assoc, Tokyo, pp 599–617 (in Japanese)

Sakai K-I, Arita M, Iyama S, Iwagami S, Okada Y, Tomita K, Hayashi S, Miyazaki Y (1978) Report on genetic resource conservation of Yanase natural sugi forest. Misc Rep Kansai For Tree Breeding Inst, Shikoku Branch, Kochi, pp 1–71 (in Japanese)

Sato K, Miyajima H (1956) On a cultivar of *Chamaecyparis obtusa*. Warm Temperate For 9:4–8 (in Japanese)

Taira H (1979) Vegetatively propagated cultivars of *Cryptomeria japonica* in Toyama Prefecture. Identification of the cultivars by means of analysis of isozyme and needle traits. Bull Toyama For Exp Stn 5:1–66

Toda R (1969) Silviculture: tree breeding. In: Japan Forest Technical Association (ed) History of forest technology, vol 3, pp 1–45 (in Japanese)

Toda R (1974) Vegetative propagation in relation to Japanese forest tree improvement, NZJ For Sci 4:410–417

Tokugawa M (1974) Historical studies on silviculture in Edo-era. Chikyu Shutsuppan KK (Tokyo) 373 pp (in Japanese)

Toyama T (1976) Historical route of sugi, *Cryptomeria japonica*. Chuokouronnsha Tokyo 215 pp (in Japanese)

Tsukada M (1980) History of sugi, *Cryptomeria japonica*: past fifteen thousand years ago, Kagaku 50:538–546 (in Japanese)

Tsukada M (1982) *Cryptomeria japonica*: Glacial refugia and late glacial and postglacial migration. Ecology 63:1091–1105

Yasue M, Ogiyama K, Ebishigawa S, Kondo K, Nishina K (1978) Diterpene hydrocarbons in *Cryptomeria japonica* needles IV. The inheritance of (–)-kaurene (+)-phyllocladene and (–)-sclarene. J Jpn For Soc 60:345–348

Yasue M, Ogiyama K, Suto S, Tsukahara H (1979) Thermatological studies on diterpene hydrocarbons in leaves of *Cryptomeria japonica*. Rep Sci Res Minist Educ Sci Cult Jpn, pp 1–52 (in Japanese)

Yasue M, Ogiyama K, Suto S, Tsukahara H, Miyahara F, Ohba K (1987) Geographical differentiation of natural cryptomeria stands analysed by diterpene hydrocarbon constitutents of individual trees. J Jpn For Soc 69:152–156

Yoshida N, Sanui T, Taketani A (1980) Biology and mechanism of injury of Japanese ceder bark midge, *resseliella odai* (Inouye). In: 16th Int Congr Entomology, Aug 3–9, Kyoto (Poster Session)

5 Strategies for Clonal Forestry with Poplars, Aspens, and Willows

L. Zsuffa[1], L. Sennerby-Forsse[2], H. Weisgerber[3], and R.B. Hall[4]

1 Introduction

Poplars (species and interspecific hybrids of sections *Aigeiros* and *Tacamahaca*), aspens (*Leuce* poplars) and willows (*Salix*) exhibit a variety of useful morphological, anatomical, and physiological characteristics. Due to their fast growth and the fact that they may be planted on many and varied sites both in and out of forests, poplars, aspens, and willows offer ample possibilities, especially in the temperate zones, of providing additional sources of raw material and thus helping to meet the continually increasing demand for wood. For example, biomass plantations of willow clones in Sweden, at a density of about 20000 cuttings per hectare, cultivated by agricultural methods, and harvested in 2–4-year cycles, produced 12–15 oven-dry metric tonnes/hectare/year of stem wood in practical scale trials. In intensively cultivated field trials, biomass yields have been reported exceeding 30 oven-dry tonnes/hectare/year (Christersson 1987 a,b). Furthermore, these fast-growing trees can enrich the forests of the higher-latitude regions, which are often lacking in tree variety, and give a new appearance to parts of the open landscape where there are few trees and bushes.

Many advantages of vegetative propagation for poplars and willows have been known for a long time. Using conventional methods of rooting cuttings, poplars and willows have been reproduced clonally in large numbers for practical cultivation in many countries. However, this method of reproduction cannot be used in the same way for all species. In particular, the aspens have few or no preformed or latent-root primordia in the shoot and it is thus more difficult to achieve practical large-scale propagation by means of cuttings. The proven advantages of clonal propagation with poplars and willows have stimulated experiments using in vitro and "rootling" techniques with aspens, and these now show positive results.

[1] Forest Genetics, Faculty of Forestry, University of Toronto, 33 Willcocks Street, Toronto, Ontario, M5S 3B3, Canada
[2] Sector of Energy Forestry, Department of Ecology and Environmental Research, Swedish University of Agricultural Sciences, P.O. Box 7072, S-75007, Uppsala, Sweden
[3] Institute of Forest Tree Breeding, Hessische Forstliche Versuchsantstalt, Prof. Oelkers Str. 6, W-3510, Hannoversch-Münden, FRG
[4] Department of Forestry, Iowa State University, Ames, IA 50011-1021, USA

Clonal Forestry II, Conservation and Application
Ed. by M.R. Ahuja and W.J. Libby

2 Poplars, Aspens, and Willows

2.1 Silvics

Poplars and aspens are interspersed throughout all forests of the Mediterranean, temperate, and colder regions of the northern hemisphere. Willows have a wider natural distribution and also occur in the southern hemisphere.

Populus and *Salix* grow over a wide range of ecological conditions. They are pioneering occupants of bare sites. Many poplar and willow species are typical of recent alluvial deposits. But *Leuce* and *Tacamahaca* poplars, and especially shrubby willow species, grow over the broadest range of ecological conditions. Aspens, especially *P. tremula* of Eurasia and *P. tremuloides* of North America form or are components of large areas of forests. Many poplars and tree-form willows grow in forest stands on flood plains of large rivers, while other poplars and aspens, and especially shrubby willows, grow in single-species and often natural single-clone clusters and rarely spread into forests.

2.2 History of Cultivation

In the International Poplar Commission's book on poplars and willows (FAO 1979), the spread of cultivation of these species is described as follows: "When the forests of the plains of the Old World were cleared and converted to agriculture, many parts of the land were left aside because no worthwhile crops could be expected from them. However, it was there that poplars and willows were successful in establishing themselves. People came to look at these areas of natural growth to satisfy their needs for timber and fuelwood. Because of the value of the trees and the proven ease of propagating them by cuttings, people started planting them near their homes and around their fields to furnish wood and provide greenery and shade."

In Mid-Asia, in the Near-East, and around the edges of the Mediterranean, poplars and willows have been closely associated with agriculture since antiquity. For many centuries, pollard willows have provided fuel and withes, while poplars provided timber and fuel. Additionally, poplar and willow leaves were used as forage and bedding for animals. Even today, from Kashmir to the Atlantic in the Old World and throughout the New World, poplars and willows mark the neighborhoods of human habitations.

The oldest poplar culture evolved using locally available trees; almost everywhere, this was various forms of *Populus nigra* (Zsuffa 1974) and *P. alba.* However, the introduction of *P. deltoides* from North America into Europe during the 17th century resulted in natural hybrids (*P. deltoides* × *nigra* = *P.* × *euramericana* syn. *P. canadensis*), which in time revolutionized ideas about poplar cultivation.

During the first years of this century, poplar cultivation was already spread in Europe with the support of industry. Voices began to be raised on all sides deploring the general state of its disorder, the ignorance of tree planters, and the ravages caused by insects and diseases. Scientists began to take interest in poplar; in France, Dode (1905); in England, Henry (1914), in Italy, Jacometti (1933); in Germany, Von Wettstein (1933); in the United States, Schreiner and Stout (1934); and in The Netherlands, Houtzagers (1937).

Willows have been used by people for a long time, especially in Europe. In the Mediterranean region, willow husbandry has been associated with agriculture since antiquity and was an established discipline in the Roman Empire during the first century as described by Pohjonen (1984). The most important species was *Salix viminalis*, which was grown mainly for basket production. Eventually, the demand for good quality baskets increased, and at the end of the 19th century, cultivation of good willow clones became necessary. Areas under willow cultivation became extensive in Europe; for instance, in France more than 100 000 ha were used for this purpose around 1900. In Germany about 400 clones were in cultivation as of 1887 (Nordberg 1928). Great efforts to develop willow husbandry in the Nordic countries were made by Makinen (1913), Nordberg (1914), Tapio (1965) (cf. Pohjonen 1984) and by Jensen (1945). However, basket-willow husbandry declined towards the middle of the 20th century, mainly because of cheap synthetic materials. More recently, the energy problems in the 1970s have focused attention on short-rotation forestry with fast-growing trees species all over the world. In the Northern Hemisphere, countries like Sweden and Finland have climatic conditions favorable to a limited number of species, and willows are the most suitable species for energy plantations.

The tree-form willows occur in natural and cultivated stands of timber trees, often in association with poplars. They can grow into large trees, 10–35 m tall, and are usually found along rivers in lowlands of warm-temperate regions, especially along the floodbanks of the Danube River in eastern Europe, the Mississippi River in North America, and in the delta of the Parana River in Argentina (FAO 1979; Stott 1984).

Various programs employing poplar and willow cultivation were brought together in 1947 with the founding of the International Poplar Commission under the aegis of the Food and Agriculture Organization (FAO). The efforts of this Commission resulted in considerable progress in the cultivation and utilization of poplars and willows, and in the knowledge of poplar and willow types and their cultivation, and led to important agreements on nomenclature, registration of clones, varietal control, and the exchange of cuttings.

2.3 Types of Plantations

The main deployment patterns within plantations include; monoclonal stands, mosaics of monoclonal blocks, mosaics of clonal rows, and intimate single-tree mixture where adjacent trees are rarely or never members of the same clone. A variety of each is practised in different countries depending on needs, tradition, tree varieties, sites, and economics. The examples that follow illustrate these practices and also facilitate discussion of clonal concepts and experiences.

2.3.1 Block Plantings

A common method for growing poplars in stands is at medium spacing (3 m to 4 m) in 20-to-30-year rotations. Pre-rooted stock is used for planting, and some form of site preparation and initial tending is practised. The stands are pruned and thinned, depending on growth rate, tree variety, and the desired tree size and quality at harvesting. Such forms of plantations are found in Nordic countries and parts of mid- and eastern Europe with aspen (*P. tremula*) seedlings and its hybrid varieties, and throughout mid-Europe and in parts of North America with euramerican and various balsam poplar hybrid clones.

A typical form of poplar cultivation with euramerican poplar clones or similar easy-to-root and fast-growing poplar types is characterized by wide spacing (5 m to 8 m) and intensive, agricultural-type management. The plantings are usually in monoclonal blocks. Large-sized pre-rooted stock is planted on well-prepared sites, and the plantations are well tended. An example of this is the well-known Italian poplar culture. It has spread, together with clones developed in Italy, throughout the world. Such poplar plantations, when established on productive agricultural soils, are sometimes co-cultivated with annual farm crops such as corn, potatoes, etc. (i.e., agroforestry) for 2 to 4 years until the crops become shaded by the poplars. In other cases, alfalfa or other forage crops are grown and grazed by cattle in such widely spaced poplar plantations.

Balsam poplars are often planted and grown within the forest, especially in mixed stands with other associated species. For example, these poplars can be used for complementation of natural regeneration of other broadleaved forest tree species. In Germany, model calculations indicate that, on comparable sites, the financial returns from sale of wood from beech/balsam poplar mixed stands may be 35–70% higher and the net profit 200–360% greater than from pure beech stands (Weisgerber 1979a). There is also a promising outlook for mixing balsam poplar in rows with conifers, e.g., spruces and firs. The balsam poplars are planted in these stands at a spacing of 30 m × 3 m and the conifers at 2 m × 2 m. The poplars and conifers can be logged together at about 80 years of age. Also, poplars can be logged earlier, at 40 to 60 years. This opens the stand and gives access for thinning. In this way, extensive areas of conifers can be opened up, thus ensuring better resistance to damage, especially by storm. Already damaged fringes of stands can be closed again in a relatively short time by planting fast-growing poplars. The above types of balsam poplar plantings can be either monoclonal or multiclonal mixtures.

Aspen species and their hybrids are especially suited for growing on forest soils. The management is usually extensive, with natural regeneration. Planting in pure stands with a 40-to-50-years rotation period is recommended. When establishing mixed-species stands aspen is interplanted with conifers. Aspen plantings have usually been of seedlings, because of difficulties in clonal propagation. However, this may change with the availability of "rootling" planting stock (Hall et al. 1989). White poplar and gray poplar (*P.* × *canescens*) are easier than aspen to root from cuttings, and are planted either as clonally propagated plants or as seedlings. Extensive management methods using natural regeneration by root suckers, with a rotation period of 40 to 60 years, are possible on soils with good water and nutrient supply. Gray poplars can be mixed with spruces and firs, and by providing shade, protect the slower growing conifers against frost and storm. On soils of suitable fertility and location, management in shorter rotation periods (10 to 15 years) is practised for the production of small timber and fibre.

A traditional form of poplar cultivation is practised in Asia, the Mid-East, and North Africa. Fastigiate poplar clones (usually of *P. nigra* and *P. alba* origin) are grown in very dense spacings (1 m) in 10-to-20-years rotations. The trees are an important source of timber for rural construction and fuel, and leafy branches are used for animal feed.

Another form of monoclonal poplar plantations has developed in North America. Unrooted cuttings of easy-to-propagate clones are used for medium-density plantations, and the production goal is usually industrial fibre. The planting sites are prepared by plowing, and the plantations are tended by cultivation and weed control. The rotation age is approximately 10 years, and the trees are either regenerated from coppice or replanted to new stock (Kaszkurewicz and White 1976). Normally, only the best plantations and clones are allowed to coppice, otherwise it is assumed that new clones will do better. In the latter case, the site must be cleared mechanically or with herbicides before replanting to new stock.

Willow stands are in many cases similar to poplar stands. However, denser plantings (2 m to 3 m) are more commonly practised than in poplars, and tending is usually less intensive. Willow cultivation is common in Europe along the Danube River system, and in Argentina in the valley of the Parana River. Willows have traditionally been planted in China, both as trees and as shrubs (Tu 1988). A variety of species and forms are planted and used for fuel, fodder, construction, and weaving throughout the world, and for land stabilization, especially in New Zealand (van Krayenoord and Hathaway 1986).

Biomass plantations are a recent concept in poplar and willow cultivation. These are very dense plantings (0.3 m to 1.5 m) of cuttings, harvested frequently (every 1 to 5 years), and subsequently regenerated by coppice growth. The blocks are monoclonal, and trees of proper clones on adequate sites and, when rightly tended, grow well and produce three or more coppice rotations. The whole aboveground portion of the trees, sometimes even the foliage, is harvested and used as a source of fiber, energy, and food. A traditional form of willow plantations with a concept similar to biomass plantations are the osiers.

2.3.2 Isolated Row Plantings

There is a variety of cultivation forms of poplars and willows with different clones in row plantings. A common form of row plantation uses Euramerican poplar clones along ditches, roads, and farm boundaries. The spacing is usually wide (4 m to 10 m) and large, rooted stock is planted in single rows, or occasionally, in double rows. The trees are cut at 20 to 40 years of age and replaced by new plantings. This form of row planting is common throughout Europe. In some countries, such as parts of Holland, France, and Italy, approximately 40 trees grow in such plantings on every hectare of farmland.

In the mid-east and North Africa, fastigiate poplars are often grown in single or multiple rows in dense (1 to 2 m) arrangements. In the Soviet Union and Canada, poplars are an important component of windbreaks. In New Zealand, row plantations of poplar check soil erosion resulting from sheep and cattle grazing on hilly land.

A traditional concept of row plantations is practised in many Old World countries with pollard Willows. The crowns of these willows are cut back periodically to the trunk and the branches are used as fuel, withes, and stakes.

2.3.3 Single Trees and Small Groups of Trees

Single trees or small groups of trees of large number of poplar, aspen and willow varieties are planted for landscaping. Some examples of the most common varieties are: *P. nigra* cv. Italica (Lombardy poplar) in temperate zones around the world, *P. nigra* var. *thevestina* in Asia and the Mediterranean region, *P.* × *euramericana* cvs. Robusta, Serotina, Regenerata, and Marilandica throughout Western and Central Europe, *P.* × *euramericana* cv. Eugenii (Carolina poplar) in North America, *P. alba* cv. Bolleana in the Mediterranean, Asia, Western Europe, and Argentina, and *S. babylonica* = *S. pendula* (weeping willow) in temperate and subtropical zones throughout the world. Many times, these trees also provide the benefit of timber for rural structures, firewood, pollards, and forage.

3 Genetic Improvement

Poplar and willow cultivars were amongst the earliest trees selected, and some of the pyramidal poplar and weeping willow varieties have been planted since recorded time (FAO 1979). Owing to the good growth of many spontaneous hybrids and the ease of cloning by rooting of cuttings, many varieties of poplars and willows had been selected and planted worldwide by the beginning of this century.

Interspecific hybridization was started in poplars (Henry 1914) and willows (Nilsson 1918) earlier than in most other forest trees. Because of the ease of crossing, interesting features, and superior growth of many hybrids, exploratory hybridization programs had mushroomed by the middle of this century. Many of the poplar and willow clones in use today resulted from such hybridization work.

Several well-planned genetic and breeding studies started after the second World War in poplar, aspen, and willow, and have resulted today in substantially improved generations of progenies and clonal selections. Examples of such studies in poplars are by Sekawin (1969) in Italy; Steenackers (1970, 1972a) in Belgium; Kopecky (1971) in Hungary; Fröhlich and Grosscurth (1973) and Mohrdiek et al. (1979) in Germany; Koster (1976) in Holland; Zsuffa (1979) in Canada; and Stettler and Heilman (1984) in the USA. Several programs also reported significant results for aspen, such as in the USA (Einspahr et al. 1963;

Benson 1982), Germany (Melchior and Seitz 1966; Hattemer and Seitz 1967; Fröhlich and Grosscurth 1973; Weisgerber 1976, 1983a, Mohrdiek 1977; Schmiedel 1982), Canada (Heimburger 1968; Zufa 1969), and Sweden (Johnsson 1976).

In *Salix*, programs in genetics and breeding of tree-form willows were started first. Examples of important achievements are in reports by Hunziker (1962) and Ragonese and Alberti (1965) from Argentina; Zufa (1967) and Krstinic (1976, 1979) from Yugoslavia; Weber (1963) from Germany; Tompa (1970) from Hungary; Hathaway (1976) from New Zealand; and May (1982) from Italy. Although selection and breeding of shrubby willows dates back to the 1930s (Stott 1984; Zsuffa et al. 1984), well-planned genetic and breeding studies were inititated only recently, as biomass plantations gained significance for production of raw materials for energy and industries. Reports of such programs and encouraging results are by Eriksson et al. (1984) and Gullberg (1988) in Sweden, and Zsuffa et al. (1984) and Zsuffa (1988) in Canada.

4 Clonal Propagation

One of the most important characteristics of many poplars and willows is their readiness to propagate vegetatively by rooting of stem cuttings. The advantage of this for practical cultivation is obvious. Besides the low cost of producing new plants, the possibility of achieving genetically identical material by very simple means is very tempting. Silviculturally, stem cuttings are used for propagating clones not only in the nursery but also in the field, where unrooted cuttings are preferred over rooted cuttings as an inexpensive and easy-to-handle planting stock.

Factors affecting rooting of cuttings are either internal or environmental, or result from the interaction of the two (Zsuffa 1976b). Internal factors influencing rooting are related to the genetic, morphological, and physiological characteristics of the clone. The most significant environmental influences on rooting are light, temperature, and moisture. Genotype-environmental interactions cause specific changes in the biology and chemistry of the clone, which affect the rooting.

The propagation ability of willows can be demonstrated by the method described by Ericsson (1982), where it is the human factor, more than the plants, that limits the speed of propagation. Through enhanced axillary shoot formation of the seedling donor plant, more than 10^9 ramets could be produced in one year under glasshouse conditions (Fig. 1). This method offers a very high rate of multiplication, and requires a smaller input of labor and technical facilities than do tissue-culture techniques. A similar technique has been employed with both aspen and cottonwood-type *Populus* for establishing research plantations (Faltonson et al. 1983).

The rooting ability of some poplar, aspen and willow species, which show variable or poor rooting from stem cuttings, can be improved by crossing with a related species that shows better rooting. Thus, most poplar species of *Aigeiros*

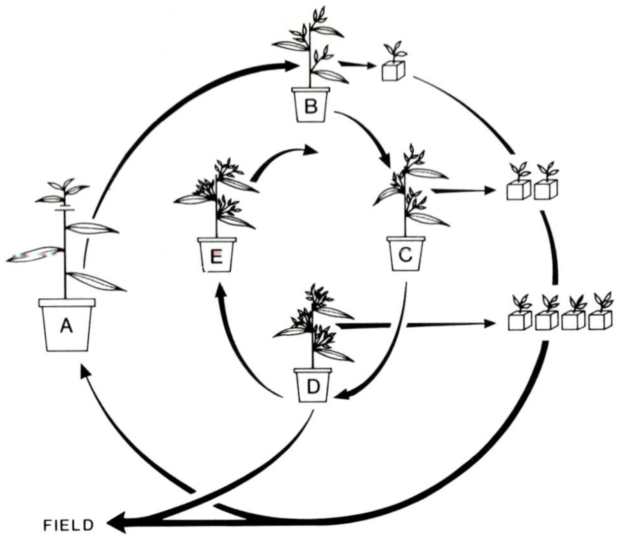

Fig. 1. Sequential steps during cutting production from axillary shoots in *Salix*. After removing the shoot apex (*A*) axillary shoots may be harvested repeatedly (*B* – *E*) for long periods. The rate of propagation is accelerated if the rooted cuttings are used as stock plants for a number of propagation cycles. (Ericsson 1982)

and *Tacamahaca* sections root with ease. However, the rooting ability of *P. deltoides* varies, and large percentages of its cuttings sometimes perish. *P. nigra* and *P. balsamifera* cuttings root more readily. These species transmit their better rooting ability to hybrids with *P. deltoides*. In fact, cuttings of *P.* × *euramericana* and *P.* × *jackii* (*P. deltoides* × *balsamifera*) clones commonly root even under less than optimum field conditions.

Within the poplars of the *Leuce* section, the European white poplar (*P. alba*) is the only species that can be propagated with dormant stem cuttings. White poplar creates fertile and vigorously growing hybrids with aspen species and at the same time transmits its rooting ability to the hybrids, as reported, for example by Heimburger (1968), for crosses of *P. alba* with *P. grandidentata*, *P. tremuloides*, *P. tremula*, *P. davidiana* and *P. sieboldii*. By backcrossing the hybrids to *P. alba* and by producing multiple hybrids of unrelated parents, such as *P.* × *canescens* × (*alba* × *grandidentata*), *P.* (*alba* × *davidiana*) × (*alba* × *grandidentata*), heterogeneous progenies with outstanding individuals were produced, in which better rooting clones could be selected (Heimburger 1968; Zufa 1969).

The poor rooting ability of aspens is a major obstacle to cloning and operational use of genetically improved stock. Most attempts to improve the rooting of cuttings have failed, especially with *P. tremuloides*, *P. tremula*, and related aspens. Therefore in recent years new ways for solving these problems were attempted by the use of in vitro techniques. Bonga and Durzan (1982) provided comprehensive information on the present state of knowledge and the importance of this technique for forestry. Research to date has concentrated mainly on *Populus tremuloides* and *P. tremula* and hybrids of *P. alba* with *P. glandulosa* or *P. grandidentata* (Winton 1971; Christie 1978; Chalupa 1981; Schmid 1981; Ahuja 1983, 1986, 1987; Barocka et al. 1985; Fröhlich and Weisgerber 1985; Chun and Hall 1984).

Poplars of section *Leuce* can be propagated by planting root segments. Cuttings approximately 10 cm in length and 0.5 to 3 cm in diameter are made from dormant root systems. When planted in nursery beds, each segment gives rise to one or more new plants, just as new sprouts form from the roots when natural stands are cut or burned. This propagation technique has been employed in the production of ornamental aspen clones and now is being studied in combination with tissue culture techniques to try to find a rapid, economic clonal deployment strategy for the hybrid aspen clones (Chun and Hall 1984; Hall et al. 1989).

Although most willows are easy to root, some species provide exceptions to this rule. In easily rooted species like *S. viminalis* and *S. dasyclados* preformed root primordia in the stems are situated mainly in the interfascicular region above lateral leaf traces. Sometimes they also occur unevenly distributed all over the internodes in dormant 1–2-year-old stems. Provided that dormant plant material is used, the primordia will develop into roots and emerge in 6–10 days in water-soaked or planted cuttings.

S. caprea is one of the willow species that is unable to form adventitious roots spontaneously. Root primordia have not been found in *S. caprea* cuttings although a hybrid between *S. caprea* × *viminalis* had these features (Fjell 1987). To root *S. caprea* cuttings, treatment with IBA is required. The initiation and development of roots then follows a different route compared to that of preformed primordia and it takes up to a month for such cuttings to root.

5 Clonal Testing

Clonal forestry is only as reliable as the clones used for plantings. The reliability of clones depends in substantial degree on the selection criteria used and on the effectiveness of testing.

Criteria often used for selection of poplar and willow clones include: good rooting ability; resistance to diseases; desired wood quality; stem form and branching habit; positive reaction to site management; and regeneration by coppicing. Some important clonal traits, such as rooting ability and stem form, can be evaluated at a juvenile age. However, other significant traits, such as growth rate, harvest volume per unit area, and disease resistance, may change with age. For example, in a clonal trial in Ontario, Canada, the ranking of clones by height changed considerably during the first 6 years of growth (Fig. 2). This meant that several years of observation were needed for proper clonal selection. Also, the fast-starter clones would be better suited for 1–3-year harvesting cycles in biomass plantations, whereas longer rotations might require different clones (Zsuffa 1975).

Clones	Ranking at age (yrs) 1	2	3	4	6
I-10/56	1(1.7m)*	10(3.7m)	10(4.6m)	4(6.1m)	11(8.5m)
I-61/59	2	6	4	11	20
Negrito de Granada	3	3	11	5	5
I-214	4	5	8	10	4
Ostia	5	8	1	20	15
Jacometti 78B	6	12	20	8	8
B-56	7	4	5	1	22
I-262	8	1	13	12	10
I-55/56	9	7	12	15	12
Eugenii**	10	9	3	13	14
Regenere Batard d'Hauterive	11	11	18	19	1
Canada Blank	12	13	6	3	19
Regenerata Harff	13	15	9	6	13
I-132/56	14	19	7	18	6
Blank de Poitou	15	18	22	9	24
Virginie de Frignicourt	16	20	2	22	3
Chopa de Sante Fe	17	2	15	2	18
Virginie de Nancy	18	14	21	7	2
Dromling	19	22	14	14	21
Robusta	20	21	19	21	9
Regenere d'Aube	21	16	16	24	7
I-45/51	22	23	24	17	23
I-455	23	24	23	16	17
Tardif de Champagne	24(1m)	17(2.5m)	17(3.1m)	23(4.4m)	16(5.9m)

* Mean height ** Clone developed from a Carolina poplar tree, selected in Orono nursery

Fig. 2. Ranking of clones according to mean heights at Huronia District, Flos Twp. (southern Ontario, lat. 44°10′N, long. 80°W) clonal test established in the spring of 1969, on Tioga loamy sand, at 10 × 10 ft. (3 × 3 m) spacing. All clones in the test are of *Populus* × *euramericana* (Dode) Guinier (= *P. deltoides* Bartr. × *P. nigra* L.) origin. The solid and broken lines indicate the changes in ranking of some clones

Many important characteristics are largely influenced by environment. One such is the growth, which depends significantly on the site. Poplar and willow clones vary in their site requirements, even if they are of the same species. In euramerican poplar clonal trials across a range of sites, some clones grew relatively well on a number of sites, whereas some performed extremely well on some sites but showed poor growth on others (Zufa and Zivanov 1966).

Clonal tests are usually organized at several levels, from the initial screening of single-copy genotypes (in progeny trials) through clonal performance testing. Libby (1987) recommended testing for clonal forestry at four levels: Level 1 – initial screening, involves seedlings, although in some cases the plants screened will be single-copy plants of tissue-culture or somatic-embryo origin. Level 2 – candidacy testing, with large numbers of clones and a small number of ramets per clone. Level 3 – clonal performance testing – more extensive testing of better candidate clones. The number of clones at this level of testing will be moderately small, and the number of ramets per clone should be large. Level 4 – compatibility trials attempt to identify compatible sets of clones that can be advantageously grown in sequenced mixtures. This level of testing will be done by advanced programs, with a relatively small number of highly successful and well known clones. Many programs practise the first three levels, while level 4 has rarely been used in clonal testing of poplars, aspens, and willows.

6 Clonal Plantation Strategies

Monoclonal planting is common in all forms of poplar and willow plantations (Zsuffa 1985). The reasons for, and practicalty of, monoclonal poplar and willow culture are: ease of propagation by rooting of cuttings; the advantages of growing and utilizing plantations of uniform trees; large variation within populations and families; and the occurrence of unusual, desired individual types. The ease of vegetative propagation alone, if dealing with relatively uniform populations and sibs containing many desired types, could have led to multiclonal cultures. However, poplar species in general, and families of interspecific hybrid poplars in particular, are well noted for large tree-to-tree variation. In fact, *P. nigra* clonal selections of fastigiate forms, such as cvs. Italica, Thevestina, and Plantierensis, which have been propagated for centuries, represent unique and rare occurrences. Old selections of euramerican poplars, such as cvs. Serotina, Regenerata, and Robusta, are also distinguished by particular features of form, growth, and site adaptation. For this reason, and because of the relatively late start of planned poplar and willow breeding programs, the number of clones used in monoclonal plantations throughout the world has remained relatively small.

The International Poplar Commission has registered only 52 clones, and, as of 1983, had only eight more clones that were candidates for registration (FAO 1983). Most of these clones are euramerican poplar hybrids. The Commission's textbook on poplars and willows (FAO 1979) described five clones of *P. nigra*, 22 clones of *P. deltoides* (in use since 1970), 44 clones of euramerican poplar (about half of them in use since 1970), and 20 clones of balsam poplars and their hybrids (eight of which have been in use since 1970). Thus, the total number of poplar clones registered is fewer than 100, and those in extensive commercial use are even fewer.

The situation with willow clones is similar, in that the total number of clones in commercial use does not exceed 100. However, many of these are osier willows; few clones are tree-form.

While osier and biomass (energy) plantations are almost always monoclonal, tree plantations of willows, especially of *Salix alba* along the Danube River, are occasionally multiclonal. This is because special selections were not made earlier within this species, which is native to the flood plain of the river, and the plantations established on similar sites are then often managed as forest stands using extensive methods common in traditional forestry.

Only a portion of registered poplar and willow clones are widely used in plantations. According to the reports of the National Poplar Commissions, submitted to the 16th Session of the International Poplar Commission, the number of clones planted on a large scale is very small in most of the countries (Chardenon 1980). In Hungary, three clones (*P.* × *euramericana* cvs. Robusta, I-214, and Marilandica) account for 81% of the plantations; in the Netherlands, euramerican poplar clones Robusta and Zealand account for 60%; and in France, Robusta and I-214 account for 70% of the plantings. In the Republic of Korea, euramerican poplar clones I-214 and I-476 are planted almost entirely on the plains, while seedlings of a single hybrid aspen family (*P. alba* × *glandulosa*) are planted in the hills and lower mountain ranges on a total of 428 000 ha. Worldwide, a few clones account for well over 50% of all poplar plantations, the extent of which can be estimated at more than 1 500 000 ha.

The situation in willow plantations with regard to the number of clones registered and planted over large areas is similar to poplar. In England, a single clone (*S. alba var. coerulea*) has been planted and used for the production of cricket bats. In osiers, as well as in tree plantings, a handful of clones is favored and used throughout the world.

The need for using more diversified clonal stock was realized in poplar and willow culture in the 1960s in conjunction with the dawn of new ideas in plantation technology, spreading of plantations to different sites, and well-planned breeding programs. The ideas on integrated use of trees (the biomass concept), and new needs and possibilities in biomass utilization (such as for composite boards, energy, food) led to studies of matching clonal characteristics to production and utilization technology. These studies showed significant clonal variation and resulted in new clonal selections and diversification.

After the Second World War, poplar plantations were attempted on new and different sites, such as forest sites, upland sites, and in different countries and climatic zones. Occasionally, traditional clones did not perform satisfactorily in such conditions and a search for new selections was initiated. At the same time, some of the well-established breeding programs have come to fruition.

All this resulted in a variety of new breeding stock with new combinations of hybrids, and a large number of new clonal selections. Although the need for diversification of clonal stock was realized and breeding efforts for this diversification were made, multiclonal concepts were considered and developed in only a few cases.

According to Libby (1987), there are two major alternatives for clonal deployment: MOMS (Mosaics of Monoclonal Stands) and WIMPs (Widespread Intimately Mixed Plantations). In practice, these two main concepts are considered and applied for clonal mixtures in poplar plantations in : (1) the use of mosaics of relatively small monoclonal blocks which are similar in size, and (2) the use of row-to-row or tree-to-tree type clonal mixtures.

An alternative strategy is available for the aspens and *Salix* species that root-sprout (Hall 1982). WIMPs can be established and managed for a long enough first rotation to force substantial tree-to-tree competition. Then, after the first harvest and regeneration from sprouts, the differential competitive ability of the individual clones on the different microsites will begin to change the plantation into more of a MOMS configuration. The advantage of this approach is that the actual site conditions are making the final selection of the clone(s) to be grown there.

The concept of mosaics of monoclonal blocks in poplar plantations was developed and implemented in Ontario, Canada (Ontario Ministry of Natural Resources 1983). According to this concept, clones are carefully matched to site, plantation system, and the production goals. Sites are planted with several, rather than one or two, different clones. Monoclonal blocks are no larger than 5.0 ha in size. Each clone in production is continuously being tested for growth and for pest resistance. As better clones become available from on-going breeding work, the poorer clones are removed from production propagation and planting. The number of clones in propagation and planting is, as a policy, more than 50 at any given time, and 5.0 to 10.0% of the clones in production are changed annually by phasing-out and new introductions.

The poplar plantation program in Ontario, as well as the breeding work in support of this program, has intensified since the late 1960s. The existing clones did not, in many cases, satisfy the needs of the program, and studies demonstrated the substantial potential for new combinations of hybrids and clones (Zsuffa 1976a, 1979). Team work and cooperation between research, technology development, and management led to the development and successful application of this multiclonal concept in Ontario (see Strobl and Evers, Chap. 12, Vol. 2, for more detail).

The use of row-to-row and tree-to-tree clonal mixtures was attempted in Holland and Germany. Kolster (1978) investigated the possibility of decreasing monoclonal risks by mixing two clones by rows in plantations established with 500 to 625 trees/ha. The results indicated that this method did not offer practical possibilities at these spacings. The growth of one clone was suppressed by the other. Also, the loss in wood production by one clone was not compensated by the increased growth of the trees of the other clone. For this reason, the mixing of rows of different clones was discouraged. The same result was expected when individual trees of different clones were mixed.

The concept of mixtures of individual trees of several clones, or "multiclonal varieties", was implemented by Weisgerber (1979b) in West Germany. Multiclonal varieties are constructed with compatible clones, and varieties with 5–20 individual clones are registered for trade in that country. Identification of the clones in the nurseries is carried out by a method that combines assessments of morphological, physiological, and biochemical characteristics in order to establish clonal compatibility.

The main reason for the introduction of multiclonal varieties in West Germany was the spread of poplar culture to forest land. Attention turned to aspens and balsam poplars, which place less demand on climate and soil than do cottonwoods and black poplars (section *Aigeiros*). While aspen stock is rarely clonal, due to difficulties in rooting of cuttings, balsam poplar stock often occurs as clones. In 1972, three balsam poplar clones comprised 54% of the market, and the proportion of balsam poplars to all poplar stock produced grew rapidly (to 70% by 1977). With the introduction of *Populus trichocarpa* provenances, efforts were made to enrich the gene pool and enable new selections. The development of multiclonal varieties is envisaged from such selections for growth in pure or mixed stands (with other deciduous and coniferous species).

Multiclonal varieties may succeed better in forest stands with species mixtures, wide spacings, thinnings, and less intensive tending practices than in intensively managed plantations. In addition, balsam poplars are more tolerant than euramerican poplars and more adapted to grow in forest stands.

Multiclonal varieties should be of compatible sets of clones that can advantageously be grown in sequenced mixtures. Clones for such mixtures should be tested in compatibility trials (Libby 1987), which demand a relatively long-term program. Multiclonal varieties, in which neighboring trees make complementary demands on the site, are theoretically more productive per unit area than monoclonal varieties, where neighboring trees make competing demands. However, the experimental evidence in support of this theory is not strong or consistent.

Multiclonal varieties, if not thoroughly tested, can create problems of maintenance and control. In addition, it will be difficult to maintain a certain predetermined clonal mixture in the stock because of problems in clonal identification and differences in the rate of rooting of cuttings of various clones. Also, it may be difficult to judge the performance of individual clones in the mixture and adjust and improve the multiclonal variety accordingly. Consequently, the preferred concept of clonal mixtures in pure and intensively managed plantations of poplars and willows is in uniform monoclonal blocks of shapes and sizes dictated by site and other needs. In his study of models of clonal plantations, Libby (1980) came to the conclusion that monoclonal plantations are frequently the best strategy, while mixtures of two or three clones are frequently the worst and rarely, or never, the best. Mixtures of a large number of clones in either multiclonal or monoclonal plantings are as safe as seedling plantations.

7 Case Studies

Several well-established programs in poplar, aspen, and willow plantations could serve as case studies in clonal strategies. We will describe here as examples the (1) Italian poplar culture, (2) hybrid poplar plantations in Eastern Ontario, Canada, (3) willow biomass energy plantations, Sweden, and (4) aspen plantings in West Germany.

7.1 The Italian Poplar Culture

Poplar cultivation in Italy has been a tradition for centuries. However, the Italian type of poplar culture, as it is known today, began its development in

Lombardy in the 1940s as the result of a cooperative government-industry effort.

In the 1930s, a fungal disease (*Venturia populina*) spread on almost all the then-common poplar clones and forced the initiation of an intensive breeding program for resistance. The Italian Association of Pulp and Paper Companies, with government support, established a Poplar Culture Institute, the first of its kind, to deal with the problem and also to improve poplar production. The task of the Institute included the whole spectrum of problems from breeding and plantation management to utilization and economics. Teamwork was employed in research, technology development and transfer, and demonstration. Nursery stock of new clones was produced and distributed at a nominal charge, and demonstrations and advice were provided to poplar growers.

New pest-resistant high-yielding clones resulted from this work, as well as a new system of poplar cultivation. The resulting yields more than doubled from 12–15 m^3/ha/yr to 30–35 m^3/ha/yr. The characteristics of this culture are complete site cultivation, planting of large 2-year-old rooted stock at a wide spacing (5 to 7 m), establishment of food-crops between the rows of poplar trees, and pruning and careful tending of trees and sites. On rotations of approximately 10 years, large trees of veneer quality are produced. The economics of this culture competed favorably with wheat production and it spread very quickly in Europe and to other continents. The new Italian clones, such as I-214, were invariably associated with this type of culture.

In addition to clone I-214, a euramerican hybrid poplar, no more than a dozen clones of similar origin resulted from this breeding program that were widely used in plantings. In the textbook of Italian poplar culture (ENCC 1964), the following clones were listed as recommended for planting: I-154, I-214, I-262, I-455, I-476, and I-45/51. Monoclonal plantings were recommended, with consideration given to matching clones with site qualities and plantation objectives, but without indicating limits to the area planted with a single clone, or considering the necessity of creating mosaics of plantings with different clones.

In the 1960s *Marsonnina brunea*, a fungus causing browning of leaves and green shoots, attacked several of the clones recommended in this program. As breeding continued, several new resistant clones were developed. However, the Italian poplar culture still relies on very few clones and large, monoclonal plantations.

7.2 Hybrid Poplar Plantations in Eastern Ontario, Canada

Poplar breeding, which has been ongoing in Ontario for the past 40 years (Zsuffa 1979), preceded and stimulated this program. In order to more fully realize the productive potential of the improved genetic stock, a major technology development program was initiated in the late1970s. This cooperative effort, by a team of researchers, field specialists, an enthusiastic operational forestry organization, and industry, led to the development of successful protocols for establishing and managing fast-growing hybrid poplar plantations on lands marginal for agriculture in eastern Ontario.

The most frequently used management system focuses on short-rotation and biomass plantations (Barkley 1983). Short-rotation plantations are established with unrooted cuttings at approximately 3 × 3 m spacing. The rotation period is about 10 years, and the objective is to produce traditional-sized pulpwood trees (approximately 20 cm in diameter, 18 m tall). Biomass plantations are established with densely planted (usually 0.3 m × 1 m) unrooted cuttings, which are harvested on 1–5 year rotations.

The management practices involve site identification, mechanical and chemical site preparation, tending, and pest control. After harvesting, the regeneration of subsequent crops is usually by coppicing from the tree stumps left in the ground, without replanting. Monoclonal plantations are established in mosaics and no single clone occupies a block larger than 5.0 hectares.

Clones are carefully matched to sites, management systems, and desired production quality. A large number of clones (at least 50) is kept in production, and includes a variety of species origins and parentages. They show site specifity and can be selected for a variety of requirements. Clone-site, spacing, and management-system trials provide the data base for matching clones to different requirements. Also, continuous breeding and selection work secure new and better clones all the time.

The program has been acclaimed in Ontario. Earlier established poplar plantations (of Carolina poplar) rarely succeeded. However, the new plantings have grown well, yielding 20–30 m^3/ha/yr, a significant achievement for the marginal sites on which they are planted. Furthermore, the hybrid poplar program has played a pioneering role in introducing intensive and cooperative forestry into Ontario.

7.3 Willow Biomass Energy Plantations, Sweden

An energy forestry program was initiated in Sweden with headquarters at the Swedish Univeristy of Agricultural Sciences, Uppsala, at the height of the energy crisis in the mid-1970s. The development of short rotation, intensively managed biomass plantations of willows has been the focus of this program (Siren et al. 1987).

In Swedish energy forestry, mainly the basket willow (*S. viminalis*) and the so-called water willow (*S. dasyclados* Wimm. in a wide sense) are used today. Both species are native to Europe. Water willow is widely distributed in Eastern Europe, from eastern Poland to the USSR, while basket willow is more widespread in western Europe. These species were imported to Sweden several hundred years ago and have become naturalized in southern and middle Sweden (Sennerby-Forsse 1986).

7.3.1 Plant Material

At present, about ten clones of *S. viminalis* and *S. dasyclados* are approved for practical cultivation in Sweden. They are planted either in mosaics of monoclonal blocks or in monoclonal plantations. The need for a broader genetic base has resulted in efforts to speed up the propagation of promising crosses in order to be able to release a number of new clones for practical-scale plantations of energy forests by the early 1990s.

A long-term breeding program of *Salix* investigates the potential for higher genetic gain in *S. viminalis* and *S. dasyclados* (Gullberg 1988). The program includes both long-term and short-term objectives. In addition to the long-term breeding population and the short-term activities, new ideas and techniques will be generated by supportive research in different disciplines related to breeding (Eriksson et al. 1984).

The testing program for collections and crosses includes Level I tests in nurseries, Level II tests on different sites, laboratory tests, and finally Level III tests carried out on different sites. Selections of clones approved for practical cultivation can be made in 5 to 8 years. Some of the most important criteria for willows in the short-rotation intensive coppicing system are: high dry-matter production, high rooting ability, high resprouting ability after repeated harvesting, frost hardiness, resistance to disease (mainly rust and canker), suitable morphology, and high biomass quality.

7.3.2 Guidelines for Cultivation

The cuttings used for plantations are produced from 1--2-year-old dormant stems, harvested from October to April. The minimum diameter of the cuttings must not be less than 8 mm and the maximum diameter should not exceed 25 mm. The cuttings are in general 20–25 cm in length. Immediately after preparation, the cuttings are placed in cold storage at $-4\,^{\circ}C$, until 2 days before planting. The cuttings should then be soaked in "tap water" for about 48 h, in order to induce growth of the preformed root primordia.

Planting with stem cuttings of willows is done on selected and prepared land. Figure 3 illustrates the sequence of planting, growing, and harvesting of energy-forestry willows. An energy plantation like this can probably last 20 to 30 years before replanting is necessary (Sennerby-Forsse 1986).

The production of woody biomass in short-rotation coppicing systems involves the following operations (Mitchell et al. 1988):

- Preparing the land in the year before planting by ploughing and herbicide treatment to kill off all the aboveground vegetation.
- Unrooted stem-cuttings are planted as early as possible in spring using a planting machine. The planting density is about 15 000–20 000 cuttings per hectare, either in twin rows or single rows.
- Herbicide treatment immediately after planting is sometimes necessary and mechanical weeding between rows may also be required (almost complete weed control is essential for successful establishment).
- Irrigation may be necessary in dry years during the first phase of establishment when the roots are developing.
- In the second year fertilizers are added once the root system develops. In the following years fertilization may be conducted; as willows respond strongly to availability of nitrogen.
- After harvesting, the woody material is either chipped at the site and

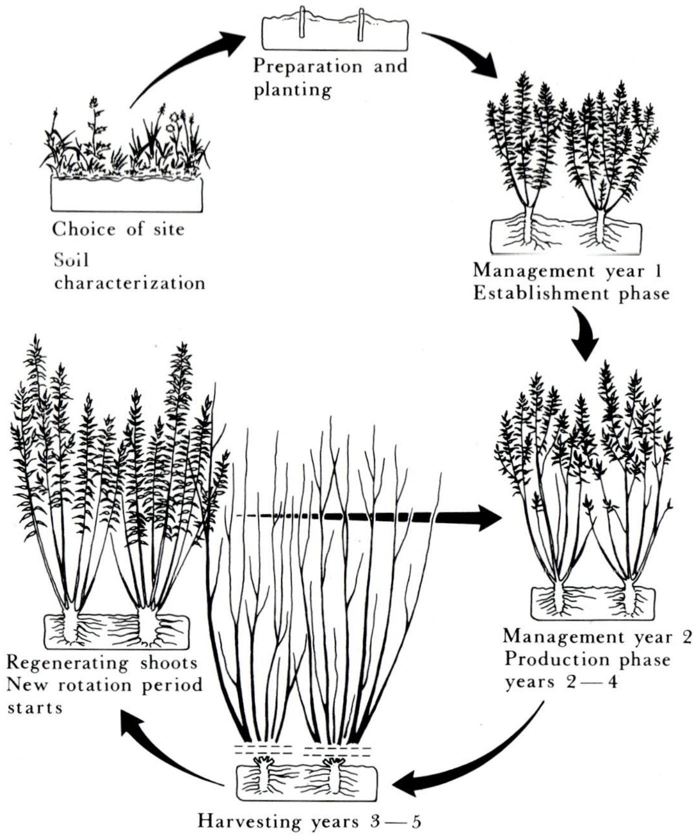

Fig. 3. Principles for energy forestry. Planting with stem-cuttings of willow is done on selected and prepared land. After 3–5 years the shoots are harvested after leaf fall in dormant conditions. A new shoot generation is developed during the following spring from the harvested stumps. These shoots are, in turn, harvested after some years and new shoots will again emerge from the stumps. The duration of an energy forestry plantation has been estimated to 20 to 30 years before re-planting is necessary. (Sennerby-Forsse 1986)

transported to a power plant for combustion or it may be chipped at the power plant. If the biomass is stored it should be kept as whips, in an aerated pile, in order to avoid energy losses and growth of injurious fungi with micro-spores and also to promote drying.

Average yields of 12–15 oven-dry-metric-tonnes/ha/yr are produced in successful biomass energy plantations.

7.4 Aspen Plantings in West Germany

Aspen is usualy planted on forest sites, often in mixtures with other hardwood species such as beech (*Fagus silvatica*), and with conifers such as spruces and firs.

Good breeding work has produced excellent aspen clones, but difficulties in vegetative propagation hindered their large scale use. Because of this, propagation by in vitro culture was studied extensively.

This program was set up in Germany in close cooperation between the Hessian Forest Research Station at Hann. Munden and the KWS Kleinwanzlebener Saatzucht A.G. at Einbeck. It was carried out in two stages: first, extensive preliminary research led in 1980 to the development of a method of optimum production and deployment. Second, since 1981, aspen clones bred and approved for trade are being propagated on a large scale and used for silvicultural purposes.

The propagation method used has been described in detail by Fröhlich (1982), Weisgerber (1983b) and Barocka et al. (1985). The timespan needed for the propagation of aspens by means of tissue culture is short: about 10 to 14 days are needed for shoot to form and about 8 to 10 days for the root, so that

BREEDING MEASURES 1955-1980*

1) Selection of 115 plus trees, carrying out of numerous controlled pollinations, in part induced mutations.
2) Testing of more than 150 interspecific progenies at 24 test sites.
3) Selection of MORE THAN 200 PLUS TREES from progenies with high-yield potential and relatively uniform growing behaviour.

VEGETATIVE PROPAGATION ± 1980

1) Until 1980 by conventional methods (rooting of cuttings); result: propagation possible on a small scale only, therefore of little significance for practical forestry.
2) Since 1980 by TISSUE CULTURE; result: LARGE SCALE PROPAGATION.

INTRODUCTION TO PRACTICAL CULTIVATION > 1981

1) Provisional grouping into MULTICLONAL VARIETIES with 20 clones in each variety.
2) In practice since 1981, after provisional approval.
3) Complementary clone examination and final grouping.
4) In wider use after final approval.

* Details refer to older test series. Different original material was used for recent research, not yet completed.

Fig. 4. The sequence of breeding, development of vegetative propagation, and clonal forestry with *Populus tremula* and *P. tremuloides*. (Fröhlich and Weisgerber 1985)

plants grown in this way can be transplanted into the greenhouse 3 or 4 weeks after the explant is taken.

In 1981, the goal of raising 50 000 plants from the triploid aspen clone Astria by means of tissue culture was achieved. After one growing season in the field, they reached heights of 187 to 257 cm, and root-collar diameters of 13 to 23 mm, with good root development.

During 1982–83, the propagation program was considerably extended. For the first time in the history of aspen silviculture, multiclonal varieties were produced. Drawing from 49 available clones, three multiclonal varieties containing various combinations of 20 clones each were deployed. In 1983, approximately 150 000 plantlings were produced.

Figure 4 gives an overview of the progress from fundamental breeding work with aspen, through the development of operational tissue-culture protocols, to the introduction of tested high-yield clones into practical forestry.

7.4.1 Multiclonal Varieties

There have been many reports on the breeding steps carried out in Hann. Munden with aspen and the results that led to the establishment of multiclonal varieties as tested reproductive material conforming to official regulations (Fröhlich and Weisgerber 1985, 1987). Currently, more than 200 clones, which originated from controlled pollinations based on earlier tests, are available for propagating and planting purposes (Fig. 4); these have a high-yield potential and relatively uniform growing behavior. In the coming years, therefore 49 aspen clones already being propagated can be complemented by a larger number of other clones that have been tested for growth behavior and pest resistance. It is planned to organize this material into different multiclonal varieties, and to allow it to be used in practical forestry in this form only. By this means it is possible to prevent the establishement of large monoclonal stands, which would be unacceptably exposed to biotic and abiotic dangers.

7.4.2 Field Plantation Experience with in vitro Propagated Material

Many of the aspens now propagated by in vitro methods could be used on a large scale for afforestation purposes. The early results and experiences are consistently positive (Schulzke 1983, 1987). The healthy and fast-growing plantlings are becoming increasingly popular in forest management. They may be used for many different purposes, especially for the establishment of mixed stands with other forest species.

7.4.3 Focus of Cultivation

Among the most favorable characteristics of aspens and their hybrids are their adaptation for cultivation on difficult forest soils. Planting in pure stands with a 40–50-year rotation period is recommended. Future treatment may be in the form of extensive management with natural regeneration, or mixing with other forest species. In this case, aspen is planted in rows in 30 × 3 m spacing, and intergrows with other species. Management in shorter rotation periods of 10–15-years at closer spacing is recommended for the production of biomass in pure stands when suitable terrain is available.

8 Experiences with Clonal Plantations

Experience has been gained from monoclonal plantations of poplars and willows, with a few clones planted on large areas, on various sites, over a relatively long period of time, and in a variety of plantings. The successes and failures of these clonal plantations with regard to performance of clones, to changes of clonal characteristics with time, and to pest problems, are discussed below.

8.1 Clonal Performance

Exceptional clonal performance was experienced with only a few clones. Before 1960, millions of ortets and clones had been screened in many breeding programs, and only several hundred clones had been retained for propagation. Mohrdiek et al. (1979) reported that, in West Germany from 1948 to 1967, 556 crosses were made and approximately 1% of the seedlings were retained for clonal testing. Very few of these several hundred tested clones equalled or surpassed the performance of then-standard clones. Thus, new clones of the quality of I-214, Robusta, and Androscoggin were and are difficult to obtain. It is surprising that these clones appeared so early and stand up to later challenges of clones from more advanced breeding programs.

Some of the exceptional clones are adapted to a large variety of sites and growing conditions. A good example is I-214, which has outperformed other clones (including local ones) in various countries of Europe, Asia, and the New World. Naturally, its ecological niche has limitations; in higher latitudes, frost limits its growth, while, in lower latitudes, it is outperformed by clones adapted to shorter days and longer growing seasons. This is the situation in the southern United States where local selections of *P. deltoides* outperform most euramerican poplars, including clone I-214.

Not only photoperiod, but site and growing conditions often define the limits of acceptable performance of clones. For example, on forest sites and in forest stands, clones of balsam poplar origin, which are less demanding and more tolerant than those of the black poplars, will outperform most clones from the latter group. In such conditions, the balsam poplar *Androscoggin* is preferred to euramerican I-214.

Unlike conifers (see Greenwood and Hutchison and Bonga and von Aderkes, Chap. 3 and 12, Vol. 1) willow and poplar clones used over a long period of time have not apparently matured and their characteristics have not changed. *P.* × *euramericana* cv. Serotina is an early selection, described in 1775.

More than 200 years later, it is still in use, and its clonal characteristics and growth pattern appear to be unchanged. The same applies to clones such as Robusta (selected in 1904) and I-214 (selected in 1940).

8.2 Pest Resistance

Compared to other tree taxa, poplars and willows are unusually susceptible to diseases, insects, and other types of biotic and abiotic damage. Therefore, resistance to pests, especially to diseases, is a prerequisite to clonal plantings of poplars and willows. Resistance breeding and screening in poplars and aspens has a long history, and has resulted, in many cases, in remarkable success. However, with some exceptions, such as *Erwinia salicis*, the watermark disease (Day 1924), the studies of pathogens are more recent and resistance breeding has just started (Hunter et al. 1988).

In poplars, clone-specific calamities were caused by pathogens such as *Melampsora* spp., *Marssonina brunea*, *Dothichiza populea*, *Xanthomonas populi*, and *Septoria musiva*. In aspens such serious pathogens are *Pollaccia radiosa* and *Hypoxylon pruinatum*. Resistance breeding and screening resulted in remarkable successes with most of these pathogens (for example, Ridé 1963; Gremmen and DeKam 1974; Steenackers 1979; Kechel 1984).

Several pathogens may pose serious threats to clonal plantations of willows, especially to dense biomass plantings. Surveys indicate that *Malampsora* rusts may create problems, as may cankers such as *glomerella* and *Fusicladium* (Hunter et al. 1988). The observed variation in susceptibility to these pathogens between species, and among clones belonging to the same species and hybrid groups, is encouraging (Royle et al. 1988). Thus resistance breeding and clonal screening for disease susceptibility in willows will most likely be successful.

Where resistant poplar and willow clones have been properly tested, clonal pest resistance has remained similar over long periods of time, unless new strains of pests have appeared, or the conditions influencing resistance have changed. Poplar clones selected in western Europe for their resistance to leaf rust (*Melampsora* spp.) have remained virtually immune for many years (Steenackers 1972b). However, some of the same clones became susceptible when moved to North America and exposed to different species and strains of leaf rust. Clones of euramerican poplar in cultivation in Lombardy (Italy) in the 1920s succumbed to a newly spreading disease (*Venturia populina*) causing spring defoliation; however, euramerican poplar clones subsequently selected for resistance to the same disease have remained resistant for over 50 years. *Marssonina brunea*, first described in North America and noted in Europe in 1958, caused serious problems in monoclonal plantations in Europe in the 1960s. More than one-half of the clones used in plantations suffered serious set-backs from the disease. Since then, new clonal selections resistant to the disease have been found and are replacing the susceptible clones. With the establishment of short-rotation coppice plantations of poplars in North America, a previously low-risk leaf-spot disease, *Septoria*, has developed into a major limiting factor, causing stem cankers that form under conditions of high inoculum and close spacings. Hybrids with *Tacamahaca* section parentage have been particularly susceptible to *Septoria* canker disease. This is causing some reorientation in breeding programs and a search for resistant *Tacamahaca* parents (Ostry and McNabb 1985). It is obvious that pests can move from surrounding natural vegetation to monocultures, that pest inocula can increase, that new strains of pests can develop, and that new pests can be imported form other countries and continents. The problems thus created are easier to overcome in short-rotation than in long-rotation monocultures, especially if the former are supported by active breeding programs. However, multiclonal plantations combined with repeated phasing-out and introduction of clones can lower the risk even further and such a strategy may be a more efficient method of general pest management.

9 Monoclonal vs. Multiclonal Plantations

The system of monoclonal plantations of poplars and willows has worked for more than a century. Most have been successful, although there have been severe set-backs in several countries due to pest problems. Monoclonal plantations are easier to manage than multiclonal plantations; fewer clones cause fewer problems in stock production, handling, and control of clone identity. The determination of clonal mixtures and matching of such clonal mixtures to sites can be a difficult and complicated task.

Clone-site trials in Ontario (OMNR 1983) demonstrated that the quality of the soil within planting sites varied significantly over a small area. Therefore, systematic analyses of the planting sites were necessary to map these changes.

Monoclonal mosaics have the advantages of lowering the risks of pest problems and of utilizing variation in performance among clones by matching these to different sites within the plantation.

Clones can also be chosen effectively for different plantation systems. For example, fastigiate and shade-tolerant clones are better adapted to dense plantings, and clones demonstrating rapid juvenile growth are better adapted to biomass plantations. The ranking of clones (for height) in trials in Ontario changed over a period of 5 years. Clones that initially grew well were surpassed by other clones, given sufficient spacing to avoid suppressive competition (Zsuffa 1975). The errors in selection for height-growth alone would have amounted to almost 50%, and the loss in volume, depending on diameter, taper, and branching, could have been more significant. This is one of the reasons why an emphasis should be placed on diameter and volume growth in making early selections (Hall 1982).

Clonal biomass qualities vary significantly, and large gains can be realized by proper matching of clones to production goals. Clonal variation in poplar biomass qualities was demonstrated in Ontario by Anderson and Zsuffa (1975, 1976, 1982, 1983) and in willow by Flower-Ellis and Olsson (1981), Sennerby-Forsse et al. (1983) and Sennerby-Forsse (1985, 1989). Very significant clonal differences (sometimes more than 50%) were observed in many qualities of wood, bark and foliage (e.g., specific gravity, fiber qualities, gross heat of combustion, nutrient content, protein concentration, and composition). The feasibility of simultaneous selection for several characteristics (such as growth rate, specific gravity, high biomass productivity, low nutrient content, etc.) was also demonstrated.

10 Clonal Strategies for the Future

10.1 Synthesis

Multiclonal plantations are a desirable alternative to existing monoclonal plantations in poplars and willows. The concept of "multiclonal varieties" in individual tree mixtures may be feasible, particularly for forest stands. The concept of mosaics of monoclonal stands, each of which is relatively small and similar in size, and in which clones are matched to soils, plantation systems, and production objectives, appears to be a reasonable choice for intensively managed plantations.

Until recently, poplar, aspen, and willow breeding programs were unable to produce a sufficient number of clones of known and desired qualities for specific sites, plantation systems, and production goals. Renewed efforts directed to the genetics and breeding of poplars and willows should provide the realization of such goals. The genetic base for breeding has been significantly broadened, large collections of native and exotic species have been assembled, and genetic variation studies are underway. Studies of genetic-environmental interactions of yield, and of genetic-architecture parameters for many traits, are yielding results. New and better-planned crosses are producing rich pools of pedigreed hybrids for selection and clonal development. Internationally coordinated programs, such as those of the International Energy Agency (IEA), the International Union of Forest Research Organizations (IUFRO), and the International Poplar Commission (IPC), facilitate good exchange of information and faster results.

Monoclonal stands are easier for the silviculturalist, the nurseryman, the designated authority for control of planting stock, and the industrial user. However, if the clonal balancing is left to the user, there is the danger that one or two clones will be favored and thus constitute the majority of plantings. In order to prevent this, each single clone should have a high and specific usability. With such sets of clones, mosaics of monoclonal plantations, consisting of clones tailored to sites, plantation systems, and production goals, will be advantageous for intensive, industrial production. On the other hand, mixtures of clones of similar characteristics, or multiclonal varieties, may be advantageous in forest stands when components, site, or their interactions vary and/or not properly known (Heybrook 1978).

10.2 Recommendations

Any clonal strategy will be only as good as the clones used for the plantings. Diseases appear to be a major limiting factor in the use of poplar and willow clones, either in monoclonal or multiclonal plantations. Therefore, sustained and even increased research efforts in resistance breeding, pathogen control mechanisms, clonal testing, and screening for resistance are recommended.

Focusing on studies that determine which clonal characteristics are important for matching clones to sites and production goals is also recommended. This knowledge is necessary for the multiclonal strategy of using mosaics of monoclonal plantations (MOMS) (Libby 1987), and also for widespread intimately mixed plantations (WIMPs). Clonal testing for the latter in compatibility trials, as recommended by Libby (1987), should be carried out.

Continuing research efforts in genetics and breeding of poplars, aspens, and willows should include a broadening of the genetic base. Given that, well-conceived breeding and selection should secure the variety of clones needed for the implementation of any reasonable clonal strategy.

Research to improve vegetative propagation, especially with aspens and those poplar and willow species that show inconsistency in rooting, is recommended.

Finally, building up national and international information systems on clones and their characterisitcs, furthering international cooperation in joint testing, and exchange of breeding and clonal stock, are important. These activities will advance the development and spread of suitable and reliable stock for future clonal plantations and strategies.

References

Ahuja MR (1983) Somatic cell differentiation and rapid clonal propagation of aspen. Silv Genet 32:131–135

Ahuja MR (1986) Aspen. In: Evans DA, Sharp WR, Ammirato PJ (eds) Handbook of plant cell culture, vol 4: Techniques and applications. Macmillan, New York, pp 626–651

Ahuja MR (1987) In vitro propagation of poplar and aspen. In: Bonga JM, Durzan DJ (eds) Cell and tissue culture in forestry, vol 3. Nijhoff, The Hague, pp 207–223

Anderson H, Zsuffa L (1975) Yield and wood quality of hybrid cottonwood grown in 2-year rotations. Ont Minist Nat Resourc, For Res Rep 101:121, 35 pp

Anderson H, Zsuffa L (1976) Farming hybrid poplar for food and fibre. An exploratory study of the seasonal above-ground biomass. Ont Minist Nat Resourc, For Res Rep 103:8 pp

Anderson H, Zsuffa L (1982) Genetically improved forest biomass energy production. In: Proc 4th Bioenergy research and development Sem, Winnipeg, Manitoba, March 29–31, pp 392–394

Anderson H, Zsuffa L (1983) Hybrid poplar energy plantations in Ontario. In: Proc ENERGO '83 Conf, Toronto, Ont, March 6–9, pp 134–145

Barkley BA (1983) A silvicultural guide for hybrid poplar in Ontario. Ont Minist Nat Resourc Publ, p 35

Barocka KH, Baus M, Lontke E, Sievert F (1985) Tissue culture as a tool for mass in vitro propogation of aspen. Z Pflanzenzücht 94:340–343

Benson MK (1982) Breeding and establishment – and promising hybrids. In: Aspen. USDA For Serv Gen Tech Rep NC-I, 88–96

Bonga JM, Durzan DJ (1982) Tissue culture in forestry, Martinus Nijhoff Publ, Den Haag Boston London

Chalupa V (1981) Clonal propagation of broad-leaved forest trees in vitro. Commun Inst For Cech 12:255–271

Chardenon J (1980) Summary of national reports on questions not the subject of analysis by subsidiary bodies of the commission. In: Int Poplar Commiss, 16th Sess, Turkey, Nov 4–8, FO:CIP/80/12: 4 pp (Mimeo)

Christersson L (1987a) High technology biomass production by *Salix* clones on a sandy soil in southern Sweden. Tree Physiol 2:261–272

Christersson L (1987b) Biomass production by irrigated and fertilized *Salix* clones. J Biomass 12:83–95

Christie CB (1978) Rapid propagation of aspen and silver poplars using tissue culture techniques. Proc Int Plant Propagators Soc 28:255–260

Chun YW, Hall RB (1984) Survival and early growth of *Populus alba* × *P. grandidentata* in vitro culture plantlets in soil. J Korean For Soc 66:1–7

Day WR (1924) The watermark disease of the cricket-bat willow (*Salix coerulea*). Oxford For 3

De Kam M (1983) The watermark disease is not transmitted with 1-year-old cuttings of *Salix alba*. Eur J For Pathol 13:212–216

Dode LA (1905) Extraits d'une monographie inédite du genre *Populus*. Mem Soc His Nat Autun 18:73 pp

Einspahr DW, van Buijtenen JP, Peckham JR (1963) Natural variation and heritability in triploid aspen. Silv Genet 12:51–58

ENCC (ed) (1964) Pioppicoltura. Ente Naz Cellul Carta Publ Roma, 130 pp

Ericsson T (1982) Propagation of *Salix* through micro-cuttings. In: N Am Poplar Counc 19th Ann Met, July 20–22. Rhinelander, Wisc

Eriksson G, Gullberg U, Kang H (1984) Breeding strategy for short rotation woody species. In: Perttu KL (ed) Ecology and management of forest biomass production systems. Swed Univ Agric Sci, Dep Ecol Environ Res Notes 15:199–216

Faltonson R, Thompson D, Gordon JC (1983) Propagation of poplar clones for controlled-environment studies. In: Methods of rapid, early selection of poplar clones for maximum yield potential: a manual of procedures. USDA For Ser Tech Rep NC-81:1–11

FAO (ed) (1979) Poplars and willows in wood production and land use. For Ser 10. FAO, Rome 328 pp

FAO (ed) (1983) Rep 31st Sess Executive committee of the international poplar commission, Casale Monferrato, It, Sept 6–8, 1982. FO:CIP/82/Rep:38 pp

Fjell I (1987) Initiation and histogenesis of adventitious roots in *Salix clones. PhD Diss, Inst Bot, Univ Stockholm*

Flower-Ellis JGK Olson L (1981) Wood density and the weight and proportion of bark in current shoots of *Salix* clones. Swed Univ Agric Sci SEP Proj Tech Rep 18:31 pp

Fröhlich HJ (1982) Fortschritte bei der vegetativen Vermehrung. Forstarchiv 53:3–9

Fröhlich HJ, Grosscurth W (1973) Züchtung, Anbau und Leistung der Pappeln. Mitt Hess Landesforstverw 10 pp

Fröhlich HJ, Weisgerber H (1985) Research on in vitro techniques within the framework of poplar breeding – results and future trends. Silv Genet 34:132–137

Fröhlich HJ, Weisgerber H (1987) Züchtungswege bis zum Aufbau von Mehrklonsorten. Forstwirtsch Centralbl 106:312–328

Gremmen J, de Kam M (1974) Research on the Poplar Canker (*Aplanobacter populi*) in the Netherlands, pt 2. Eur J For Pathol 4:175–181

Gullberg U (1988) Present state of the Swedish breeding program aiming at making an energy crop of *Salix* species. In: Proc IEA/BA Willow breeding Symp, Uppsala, Aug 31-Sept. 1, 1987. Dep For Gen Univ Agric Sci Sw Res Note 41

Hall RB (1982) Breeding trees for intensive culture. In: Proc IUFRO Joint Meet Working Parties Genetics about breeding strategies including multiclonal varieties, Escherode, FRG, pp 182–193

Hall RB, Colletti JP, Schultz RC, Faltonson RR, Kolison SH Jr, Hanna RD, Hillson TD, Morrison JW (1989) Commercial-scale vegetative propagation of aspens. In: Proc Aspen Symp, 25–27 July, Duluth, MN. USDA For Serv Genet Tech Rep NC-140:211–219

Hathaway RL (1976) Breeding willows for soil conservation and river control. Soil Water Wellington 12. 6:2

Hattemer H, Seitz FW (1967) Einige Ergebnisse von Testanbauten mit Aspenhybriden. 2 Kreuzungen der Jahre 1953 und 1958. Silv Genet 16:6–13

Heimburger C (1968) Poplar breeding in Canada. In: Maini JS, Cayford JH (eds) Growth and utilization of poplars in Canada. Can Dep For Rural Dev, For Branch Publ 1205:88–100

Henry A (1914) Note on *Populus generosa* Gardeners Chronicle 1914:257–258

Heybrook HM (1978) Primary considerations; multiplication and genetic diversity. Unasylva 30 (119–120):27–33

Houtzagers G (1937) Het Geslacht *Populus* in verband met zijn Beteekenis woor de Houtteelt. Veeman & Zonen, Wageningen, 291 pp

Hunter T, Royle DJ, Stott KG (1988) Disease of *Salix* biomass plantations – results of an international survey in 1987. In: Proc IEA/BA Task II Worksh Biotechnology development, Uppsala, Sw, Sept 1, 1987. Joint IEA/Univ Toronto Centre Plant Biotechnol Rep, Int Rep 88:1, 76:37–48

Hunziker JH (1962) The origin of the hybrid triploid willows cultivated in Argentina. Silv Genet 11(5/6):151–153

Jacometti G (1933) Ist Miglioramento Pioppo, Torino S.A.F. Publication, Roma, Italy

Jensen C (1945) Bogen om pil. Jorgensen & Bogtrykkeri. Kopenhamn

Johnsson H (1976) Das Produktionspotential der Hybridaspe (*Populus tremula* × *tremuloides*) in Südschweden. Holzzucht 30:19–22

Kaszkurewicz A, White G (1976) Planting techniques and establishement of cottonwood plantations – a review. In: Proc Symp Eastern cottonwood and related species, Poplar, Counc N Am. Louisiana State Univ Press, Baton Rouge, pp 370–378

Kechel HG (1984) Untersuchungen űber die Resistenz von Pappeln gegenűber dem Erreger des Pappelkrebses, *Xanthomonas populi* subsp. *populi* (Ridé) Ridé und Ridé. Schr Forschungsinst Schnellwachsende Baumarten, Vol 3, Univ Hannover-Münden

Kolster HW (1978) Menging van klonèn in populierenbeplantingen. Populier 15(1,2):27–32

Kopecky F (1971) History, present state and future tasks of poplar breeding in Hungary. Proc Hung Forest Res Inst 67(2):29–53

Koster R (1976) Aussichten der Pappelzüchtung in Holland. Holzzucht 30, 2/4:26–28

Krstinic A (1976) Tree breeding and silviculture of willows in Yugoslavia. In: Proc 16th IUFRO Congr, Oslo, Div 2, Proj Group 2.02:39–44

Krstinic A (1979) *S. alba*. In: FAO technical consultation on fast-growing plantation broadleaved trees for Mediterranean and temperate zones, vol 1, Lisbon. FAO, Rome, pp 383–400

Libby WJ (1980) What is the safe number of clones per plantation? In: Heybroek HM, Stephan BR, von Weissenberg K (eds) Resistance to diseases and pests in forest trees. Purdoc, Wageningen, pp 342–360

Libby WJ (1987) Testing for clonal forestry. Annales Forestales 13/1–2:69–74

Makinen L (1913) Pajun viljelemisesta koritoita varten. Tapio. Suomen Metsanhoitoyhdisty Tapion Julkaisema Aikakauskirja 6:334–337, 349–353

May S (1982) Willows for wood production. FAO Int Poplar Comm Ad Hoc Committee on Breeding, Casale Monferrato, It, p 6

Melchior GH, Seitz FW (1966) Einige Ergebnisse bei Testanbauten mit Aspenhybriden. I. Silv Genet 15:127–133

Mitchell CP, Sennerby-Forsse L, Zsuffa L (1988) Biomass qualities and potential for genetic improvement in poplars and willows. J Biomass 17:21–37

Mohrdiek O (1977) Hybridaspen für forstliche Grenzertragsböden. Forstarchiv 48:158–163

Mohrdiek O, Muhs HJ, Melchior GH (1979) Some results of the breeding program with poplar of the Aigeiros and Tacamahaca sections at Schmalenbeck. In: Proc Joint Meet IUFRO Working Parties S2-02-10 and S2-03-07, Poplars in France (Orleans) and Belgium (Geraardsbergen) pp 127–142

Nilsson H (1918) Experimentelle Studien über Variabilität, Spaltung, Artbildung und Evolution in der Gattung *Salix*. Lunds Univ Arsskifit, NF Aud 2, Bol 14, 28:144pp

Nordberg S (1914) Pajuvesametsaviljelksesta. Tapio. Suomen Metsanhoitoydistys Tapion Julkaisema Aikakauskirja 7:353–358

Nordberg S (1928) Vertaileva katsaus pajun viljelykseen ja sen edellytyksiin ulkomailla ja Suomessa. Dtsch Ref: Die Weidenkultur und ihre Voraussetzungen im Ausland und Suomi (Finnland). Silv Fenn 9:1–63

OMNR – Ontario Ministry of Natural Resources (ed) (1983) New forests in eastern Ontario. Hybrid Poplar Sci Technol Ser 1:336pp

Ostry ME, McNabb HS Jr (1985) Susceptibility of *Populus* species and hybrids to disease in the North Central United States. Plant Disease 69:755–757

Pohjonen V (1984) Biomass production with willows – what did we know before the energy crisis? In: Perttu KL (ed) Ecology and management of forest biomass production systems. Swed Univ Agric Sci, Dep Ecol Environ Res Notes 15:563–587

Ragonese AE, Alberti F (1965) Nuevos sauces hibridos forestales obtenidos en la Republica

Argentina [*Salix babylonica* (female) × *S. alba* (male) cvs. A-131–25 e A-131–27]. Idia, Supl For 2. 16:65–74

Ridé M (1963) Donnees actuelles sur le chancre bactérien du peuplier provoqué par *Aplanobacterium populi*. FAO/CIP/MAL 23 Casale Monferrato, It, pp 18–43

Royle DJ, Hunter T, Stott KG (1988) Disease of *Salix* biomass plantations: effects on yields, factors affecting clonal yields and research requirements. In: Proc IEA/BA Task II Worksh Biotechnology development, Uppsala, Sw, Sept 1. Joint IEA/Univ Toronto Centre Plant Biotechnol Rep, Int Rep 88:1, 76, pp 49–51

Schmid A (1981) Pour la promotion du peuplier tremble. Mitt Schweiz Pappel-Arbeitsgemeinschaft 32:2–8

Schmiedel H (1982) Aspenanbau in Holzzuchtplantagen. Sozialist Forstwirtsch 32:338–340

Schreiner EJ, Stout AB (1934) Description of ten new hybrids. Bull Torrey Bot Club 61: 449–460

Schulzke R (1983) Uber die Probleme bei der Verwendung von Aspen. Holzzucht 37:10–11

Schulzke R (1987) Erste Erfahrungen mit Aspen aus In-vitro-Vermehrung im Forstbetrieb. Holzzucht 41:34–36

Sekawin M (1969) Selection et amelioration. Inst Peuplier, Turquie. Rapp Tech FAO/SF:41/Tur-6, vol II:131–162

Sennerby-Forsse L (1985) Clonal variation of wood specific gravity moisture content and stem bark percentage in 1-year-old shoots of 20 fast-growing *Salix* clones. Can J For Res 15:531–534

Sennerby-Forsse L (ed) (1986) Handbook of energy forestry. Swed Univ Agric Sci, Dept Ecol Environm Res Dect Energ For

Sennerby-Forsse L (1989) Wood structure and quality in natural stands of *Salix caprea* and *Salix pentandra*; Studia Forestalia Swecica no. 182: 17 pp

Sennerby-Forsse L, Siren G, Lestander T (1983) Results from the first preliminary test with short rotation willow clones. Swed Univ Agric Sci SEF. PROJ Tech Rep 30:37 pp

Siren G, Sennerby-Forsse L, Ledin S (1987) Energy plantations – short rotation forestry in Sweden. In: Hall DO, Overend RP (eds) Biomass. John Wiley & Sons, New York 504 pp:119–143

Steenackers V (1970) La populiculture acutelle. Bull Soc R For Belg, Publ.:38 pp

Steenackers V (1972a) Breeding poplars resistant to various diseases. In: USDA (ed) Biology of rust resistance in forest trees. Proc NATO-IUFRO Adv Study Inst, Aug 17–24, 1969. USDA For Serv Misc Publ 1221, pp 599–607

Steenackers V (1972b) The state of knowledge in breeding rust resistant poplars. In: USDA (ed) Biology of rust resistance in forest trees. Proc NATO-IUFRO Adv Study Inst, Aug 17–24, 1969. USDA For Serv Misc Publ 1221, pp 419–430

Steenackers V (1979) Rapport d'activité 1978 de la station de populiculture, Geraardsbergen Station de Populiculture, Geraardsbergen, Belgium Publ. 199–209

Stettler RF, Heilman PE (1984) Short rotation poplar culture in the Pacific northwest: components of a genetic program. In: Perttu K (ed) Ecology and management of forest biomass production systems. Swed Univ Agric Sci, Dep Ecol Environ Res Notes 15:217–232

Stott KG (1984) Improving the biomass potential of willow by selection and breeding. In: Perttu KL (ed) Ecology and management of forest biomass production systems. Swed Univ Agric Sci, Dep Ecol Environm Res Notes 15:233–260

Tapio E (1965) Pajunviljely ja sen mahdollisuudet Suomessa. Konekirjaite Helsingin yliopiston kasvinviljelytieteen laitoksella, 109 pp (mimeographed)

Tompa K (1970) Breeding and production of willows in Hungary. FAO Hung Natl Committ Forestry and wood industry, Experts Committee Spec Pap 6:25

Tu Zh (1988) Willow biomass and its utilization. In: Joint IUFRO/IPC Worksh Biomass Research, Beijing, China, Sept 5–7, (mimeographed)

Van Krayenoord CWS, Hathaway RL (eds) (1986) Plant material handbook for soil conservation, vols 1,2. Water and Soil miscellaneous publication 93. National Water and Soil conservation Authority Publ. Wellington

Von Wettstein W (1933) Die Kreuzungsmethode und die Beschreibung von F_1-Bastarden bei *Populus*. Pflanzenzüchtung 1933:597–626

Weber E (1963) Genetische, pflanzenzüchterische und baumschultechnische Untersuchungen an Baumweiden, vol 2. Inst Forstl Samenk Pflanzenzücht München, 120 pp

Weisgerber H (1976) Untersuchungen über das Wuchsverhalten von Aspenhybriden auf ungünstigen Waldstandorten in Deutschland. Holzzucht 30:48–53

Weisgerber H (1979a) Holzmassen- und Geldertäge in Buchen-Balsampappel-Mischbeständen. Holzzucht 33:1–6

Weisgerber H (1979b) The importance of the international *Populus trichocarpa* provenance trial for breeding and cultivation of balsam poplars in the Federal Rebublic of Germany. In: Proc Joint Meet IUFRO Meet Working Parties S2-02-10 and S2-03-07 Poplars in France (Orleans) and Belgium, (Geraardsbergen) Sept 17–22, pp 215–225

Weisgerber H (1983a) Wuchsverhalten und Anbaumöglichkeiten einiger neu zum Handel zugelassener Balsampappeln und Aspen. Holzzucht 37:2–11

Weisgerber H (1983b) Forstpflanzenzüchtung. Aufgaben, Ergebnisse und Ziele von Züchtungsarbeiten mit Waldbäumen in Hessen. Mitt Hess Landesforstverw 19

Winton LL (1971) Tissue culture propagation of European aspen. For Sci 17:348–350

Zsuffa L (1974) The genetics of *Populus nigra* L. Annal For Zagreb VI. 2:53 pp

Zsuffa L (1975) Some problems of hybrid poplar selections and management in Ontario. For Chron 51(6):240–242

Zsuffa L (1976a) Grundlagen und Aussichten der Pappelzüchtung in Ontario, Kanada. Holzzucht 30(2–4):37–40

Zsuffa L (1976b) Vegetative propagation of cottonwood by rooting cuttings. Invited pap. In: Assoc Cottonwood Symp, Greenville, Miss, pp 99–108

Zsuffa L (1979) A breeding program for short rotation poplar biomass production in Ontario, Canada. In: Proc Joint Meet IUFRO Working Parties S2-02-10 and S2-03-07 Poplars in France and Belgium, Sept 17–22, pp 247–259

Zsuffa L (1985) Concepts and experiences in clonal plantations of hardwoods. In: Proc 19th Meet CTIA, pt 2. Symp Clonal forestry, Toronto, Aug 22–26, 1983, pp 12–25

Zsuffa L (1988) A review of progress in selecting and breeding North American *Salix* species for energy plantations at Fac For, Univ Toronto. In: Proc IEA/BA Willow breeding Symp, Uppsala, Aug 31–Sept 1, 1987. Dep For Genet Univ Agric Sci Sweden Res Note 41:41–51

Zsuffa L, Mosseler AJ, Raj Y (1984) Prospects for interspecific hybridization in willow for biomass production. In: Perttu KL (ed) Ecology and management of forest biomass production systems. Swed Univ Agric Sci, Dep Ecol Environ Res Notes 15:233–260

Zufa L (1967) Observations et experiences acquises dans l'hybridation controlé des saules à haute tige. Topola 61–64:37–53

Zufa L (1969) Poplar breeding in Canada. For Chron 45:402–408

Zufa L, Zivanov N (1966) Indications of significant and specific correlation between the poplar clones and soil types. Sumarski List 90(1–2):137–148

6 Strategies for Clonal Forestry with Norway Spruce

B.G. Bentzer[1]

1 Introduction

Norway spruce [*Picea abies* (L.) Karst.] is the most extensively used forest-tree species in Europe. Recently, increased interest in its use in northeastern USA and eastern Canada has been expressed; in some areas of this region, its growth and other qualities exceed the indigenous species (Fowler 1979). Norway spruce is fast-growing and healthy, with diversified uses for pulp as well as for saw-timber. Its rotation times are between 50 years in central Europe to 150 years in northern Scandinavia. Substantial programs to improve growth and quality of Norway spruce were initiated around 1920 in Scandinavia. Their primary focus was on provenance research, with many provenance tests being established between 1917 and 1970 (Dietrichson 1979). These studies showed that moving provenances from south to north and/or from east to west generally resulted in per-unit-area gains in volume production. However, the volume gains were sometimes correlated with larger risks, associated with both loss of hardiness and lower stem quality. Nevertheless, as of 1990, most of the gains that have been realized result from appropriate research-based provenance transfers.

Substantial breeding work began in the 1950s, with the selection of plus-trees and the establishment of first-generation grafted seed-orchards (SO). It is interesting that these pioneering efforts with Norway spruce led to what is now considered classical tree improvement; namely selection, testing, and propagation by seed in rogued open-pollinated seed-orchards (Zobel and Talbert 1984). However, for Norway spruce, the seed-orchard approach has not yet been as successful as originally anticipated. Today, in Swedish orchards, average seed-crops of only 5 kg/ha/yr are expected (5 kg produces about 600000 seedlings), and these only from orchards more than 20 years after establishment (Samuelson 1979). Several problems have slowed the pace of improved seed production. Norway spruce is notoriously slow to start flowering. A lag of 15–20 years between grafting and commencement of production-level flowering is normal. Then, once the grafts are of sufficient size or maturity to flower abundantly, the flowering is irregular among clones, as well as between years.

Judging from recent seed-orchard programs (Table 1), it seems that they will not supply sufficient numbers of improved seedlings to meet reforestation needs for a long time – if ever. (The Norwegian SO program has been comparatively

[1]Hilleshög AB, Box 302. S-261 23 Landskrona, Sweden

Clonal Forestry II, Conservation and Application
Ed. by M.R. Ahuja and W.J. Libby

Table 1. Seed orchard (SO) programs with Norway spruce operated by the various organizations

Program	Total annual demand for Spruce plants (mil.)	SO area (ha)	Accumulated production (kg)[a] until 1987	Estimated SO expansion	Time when SO production will meet need
Niedersachsen (G)	10	20[d]	5	No	?
Denmark (D)	7	40[e]	85	No[i]	2030
Hilleshög (H)[b]	180[c]	292[f]	2500	Large[j]	2020
Norway (N)	50	200[g]	6000	Modest	Never
Finland (F)	60	300[h]	100	No	Never

[a] One kilo seed is equivalent to ca. 125 000 seedlings.
[b] Most seed orchards in Sweden are established and operated by the Institute for Forest Improvement.
[c] Total Swedish demand.
[d] 19 ha more than 20 years old; 3 ha between 10 and 20 years.
[e] Thirteen ha more than 18 years old; 4 ha between 18 and 8 years old; and 23 ha less than 8 years old.
[f] Mean age of 20 years.
[g] 160 ha more than 18 years old and 40 ha less than 8 years old.
[h] Mean age of 20 years.
[i] A grafted SO will be established in a greenhouse within the next 5 years.
[j] Ten ha per year until 2000.

successful, but most of the seeds currently being produced are bred for high elevations, where the demand is limited.) The disappointing performance of the SO approach for genetic improvement and seed production resulted in a search for alternative methods. Vegetative propagation (VP) by rooted cuttings (producing stecklings) appeared to be a possibility. A paper by Kleinschmit et al. (1973), in which they described a way to select and propagate from young (4 years old) seedlings, became a starting point for several applied programs aimed at the production of stecklings for reforestation. This chapter will briefly describe and compare the recent status of five of these programs: (1) the state program of Niedersachsen in Germany (G), headed by Jochen Kleinschmit; (2) the state program of Denmark (D), headed by Hans Roulund; (3) the industrial program of Hilleshög AB in Sweden (H), headed by Bengt Bentzer; (4) the state program of Norway (N), headed by Tore Skröppa; and (5) the semi-state program of Finland (F), headed by Jouni Mikola and Martti Lepistö.

For the H program, VP is the only method of deployment to production forests. The other four also have seed-orchards and are engaged in traditional improvement work such as seed-stand selections, provenance tests, etc. In Sweden, an additional VP program for Norway spruce, headed by Martin Werner, is operated by the semi-state Institute for Forest Improvement (Werner 1977; Werner and Pettersson 1981). This organization has overall responsibility for forest-tree breeding and related research and development in Sweden. As of 1990, the five compared programs were between 16 (H) and 39 (G) years old. The D, N, and F programs all began in the 1960s. In all cases, smaller trials with VP had been made prior to the establishment of the operational programs described below.

Breeding efforts to date have resulted in various numbers and percentages of selected parent trees being progeny-tested (Table 2). With the prevailing current

Table 2. Size of breeding population and stage of progeny tests for five different programs of Norway spruce improvement

Program[a]	No. of genotypes in breeding pop.	No. of breeding zones	Percent of breeding pop. under progeny test	Percent of breeding pop. with breeding values
G	550	3	100	10
D	1030[b]	2	8	8
H	2000[c]	5	0[d]	0
N	2000	11	100	25
F	1400	7	30	—

[a] Program codes explained in text.
[b] An additional 1700 genotypes have recently been selected in nurseries at an age of 4 years.
[c] Genotypes in Hilleshög's breeding population only; not more than 17 years old.
[d] All genotypes in the breeding population are in field clone tests.

reluctance to establish more of the traditional outdoor grafted SOs, VP seems to be a compelling alternative for the effective use of progeny-test information and of limited numbers of pedigreed progeny. VP can be used in two ways: to propagate outstanding families without keeping track of individual genotypes (vegetative multiplication – VM); or to propagate, with tracking, individual genotypes selected in such families, or from other sources such as tested provenances (clonal forestry – CF).

2 Why Rooted Cuttings?

The rationales for using VP, as stated by the leaders of the five programs, are not identical (Table 3). Although problems with Norway spruce SOs are a general phenomenon, only G states that this is the primary reason for adopting VP. The capture of favorable nonadditive genetic variation specific to particular genotypes for CF, or of even a small part of it in a particular specific combination of parents for VM, are ranked high in all programs. The control of genetic variation leading to genetic homogeneity (this does not imply loss of genetic variation, but rather its controlled deployment) in plant production (nursery, tissue-culture lab, or somatic-embryo factory) (D) as well as in the forest (N, F) are also regarded as important factors. New information from clonal tests and/or from operational plantations may suggest changes in a deployed set of clones or of VM families. Such changes can be accomplished in 1–4 years. This gives a high degree of flexibility and predictability to systems using VP, a factor that was appreciated by G and H. Only F stressed the time difference between realizing genetic gain with stecklings compared to seedlings, and thus ranked the quick capture of gains as their most important consideration. This priority may be included in the top G priority, i.e., adopting VP to be able to realize any gains at all beyond those from provenance movement.

Table 3. The three most important arguments for the use of Norway spruce rooted cuttings as ranked by five programs

Program[a]	Priority	Arguments
G	1.	No improved seed from seed orchards available
	2.	Realization of extra gain, e.g., caused by nonadditive gene actions
	3.	Flexibility
D	1.	Realization of extra gain
	2.	Homogeneity
	3.	Reliable plant production of improved material
H	1.	Realization of extra gain
	2.	Climatic adaptation
	3.	Flexibility
N	1.	Realization of extra gain in growth and climatic adaptation
	2.	Homogeneity
	3.	More robust plants
F	1.	Quick utilization of gain
	2.	Realization of extra gain
	3.	Homogeneity and genotype/site matching

[a] Program codes explained in text.

3 Description of the Programs

3.1 Niedersachsen (G)

The breeding in support of G is a joint program among the states of Niedersachsen, Schleswig-Holstein, Nordrhein-Westfalen and Rheinland-Pfalz. The Norway spruce forests of this region are subdivided into three breeding zones: (1) low elevation; (2) low elevation with risk of snowbreakage; and (3) high elevation. Clones have thus far only been selected for zones (1) and (2).

The first clonal test was planted in 1947. Four years later, a test was established comparing seedlings and stecklings from the same full-sib families. Between 1951 and 1968, many VP experiments were conducted, resulting in a practically oriented selection and VP program that was launched in 1968 (Kleinschmit et al. 1973), still operating with minor modifications.

The base for the program is the annual selection, mainly in commercial nurseries, of about 4000 seedlings from approved seed sources of appropriate provenances. The initial selection is made for height only, and the intensity of this selection is about 0.00033 (i.e., 4000 selected from 12 000 000 candidates, or 1 in 3000). Thus far, only a relatively small number of seedlings has been selected from families produced by the breeding program. The selected seedlings are cloned by rooting cuttings, and their ramets are then tested as contiguous clones under nursery conditions for 2 years. As a result of this first candidacy screening, about 2/3 of the clones are culled, and about 1300 are serially repropagated for field testing and for the nursery; the nursery ramets constitute the base for further serial propagation for reforestation. The field tests are evaluated after 3 years, and the number of clones maintained in the nursery is culled by about one-third to about 860. The next field-test evaluation is made after 6 more years, and a one-quarter culling is based on these results, reducing the nursery base to about 650 clones. This procedure repeats at intervals of 6 years until the field tests are 21 years old. By that time the clones have been culled to about 10% of the original number, i.e., to about 400. Field testing is used to evaluate not only height growth but, increasing with age of the test, also other traits such as volume, site adaptation, stability over different locations, pest tolerance, etc.

Clones maintained in the nursery are serially propagated at intervals of 3 years. Cuttings are taken from 3-year-old stecklings in the nursery, and the donors are then used for reforestation. This procedure is repeated until the clone is culled.

It is anticipated that the best clones in the field tests will be entered in the breeding program. Meanwhile, open-pollinated seeds from such outstanding clones can augment seed-orchard seeds. The population of clones now in tests (about 25 000) is regarded as being superior to the population of grafted plus-trees now in the SOs; thus the pollen from neighbors in the clonal tests, even if the pollen-donating clone has been culled, is probably better than wild pollen, and similar to SO pollen.

Since 1974, as an addition to the 1968 program, a substantial effort has been put into inter-species and inter-provenance hybridization. The short-term goal is to screen families from many different combinations and identify those with outstanding hybrid vigor. (A long-term goal is to increase the genetic base of the breeding program by introgressive hybridization.) Outstanding individuals from promising families are then included in normal clonal testing. Currently, 34 different species of *Picea* are included in the program. The most successful families so far have been crosses of Norway spruce × Sitka spruce [*P. sitchensis* (Bong.) Carr.] and of Sitka spruce × Serbian spruce [*P. omorika* (Pančić) Purkyne]. Successful production of stecklings and deployment of tested clones will not have a short-term impact on deployable seedling production from the breeding program. However, it will catalyze research into the economics of hybrid production and propagation, extended test areas, less strict regulations, and probably result in more efforts directed to species hybridization.

In G, the VP program proceeds based on its own merits, and does not constitute an integrated part of the regional seed-orchard program. At present, VM of families is not allowed in Germany, and little attention is paid to specific combining ability for particular families. Combined family/clonal testing is not practised. The size and composition of the regional breeding population will, however, be affected by the Niedersachsen VP program, since the clones in the VP program augment the breeding-program clones selected by classical procedures. The total breeding population is considered to be of sufficient size so that long-term improvement can be anticipated and inbreeding can be avoided.

3.1.1 Possible Drawbacks with Clones

The major concern for the program is the maturation of clones. Stecklings derived from mature trees are known to display a relatively slow growth rate, plagiotropic growth habit, and a poor root system. Although this is a concern, the periodic selection of new clones will supply the program with sufficiently juvenile material both within and across breeding cycles to diminish this problem. In other words, in the Niedersachsen program, it is and will continue to be possible to replace maturing clones with repeated "waves" of new clonal selections. The oldest clones are now in their seventh serial-propagation cycle, and most are still sufficiently juvenile for commercial propagation. Many of these clones will be useful beyond a chronological age of 20 years, provided that the cutting-donors are serially propagated or hedged appropriately.

3.1.2 Production

Cuttings are harvested from donor plants in late winter and rooted in the spring. After rooting in plastic houses, they are transplanted into an outdoor nursery in the fall and left there for two entire growing seasons before the stecklings are lifted and delivered. Reported and anticipated production numbers from 1970 to 2000 are given in Table 4, and the rooting percentages achieved in three decades are given in Table 5. The drop in production in 1987 is a reflection of a general decrease in the planting of Norway spruce in Germany, primarily for environmental policy reasons. However, the ratio of planted stecklings to seedlings increased between 1980 and 1987.

The most apparent production drawback is the labor cost, especially for the insertion and lifting of cuttings into and from the rooting beds. Mechanization may be developed for these phases. Frost heaving of newly transplanted rooted cuttings is another unsolved problem. All stecklings are grown and delivered as bare-root plants.

Vegetative multiplication may be a production option in the future, but only if maturation turns out to be an unsurmountable obstacle. The lower gains that are expected from VM compared to CF make the former less appreciated.

Only 70% rooting success is judged to be sufficient for large-scale production. The price of stecklings relative to seedlings of similar size and physiological quality is given in Table 6. There does not seem to be a reluctance by German foresters to make the 50% additional investment in the price of the stecklings.

Table 4. Produced and anticipated production of Norway spruce rooted cuttings in five different programs (millions)

Program[a]	1970	1980	1987	1990	2000
G	0.1	1.0	0.7	1.4	2.0
D	0.01	0.15	0.65	1.0	–
H[b]	–	1.5	2.5	4.0	20.0
N	0.01	0.01	0.8	2.0	8.0
F	0.05	0.1	0.2	0.2	0.7

[a] Program codes explained in text.
[b] In Sweden, also other organizations produce rooted cuttings the number of which is not included in these figures.

Table 5. Average rooting success of Norway spruce cuttings in five different programs (% rooted)

Program[a]	1970	1980	1987
G	70	80–95	80–95
D	–	90	50–70
H	–	90	95
N	70	80	95
F	80	85	60

[a] Program codes explained in text.

Table 6. Unit price for rooted cuttings and comparable seedlings of Norway spruce in five programs (local currency, 1987)

Program[a]	Seedling	Cutting	Cutting/seedling
G	0.4	0.6	1.50
D	1.7	2.5	1.47
H	1.2	1.95	1.63
N	1.0	1.5	1.50
F[b]	0.75	1.0	1.33

[a] Program codes explained in text.
[b] All production of rooted cuttings has so far been partly or completely financed by the state.

3.1.3 Gains

An average gain in early height-growth of 10–20% is expected, based on field tests, and available data indicate that the stecklings survive slightly better than do seedlings after planting, probably due to successful clonal selection for health and adaptedness. Gains in volume, stability, and pest resistance are expected as information from the older test results in refinement of the sets of deployed clones.

3.1.4 Utilization of Clones

The large number of clones in various stages of test makes it unlikely that more than 20000 ramets of any single clone will be used in reforestation. The general opinion in Germany is that no clone should be planted on more than 1% of the total plantation area. German regulations state that a minimum of 500 individually tested clones should be used in any deployed clone mixture (Muhs, Chap. 13, Vol. 2). This standard is currently exceeded severalfold, with 3000 to 5000 clones in the two mixtures now deployed. One mixture is for each of the two zones with the most productive site indexes, where their expected higher growth and quality will be best utilized.

3.1.5 Research

Efforts are concentrated on various possible genetical, physiological, and environmental causes of maturation. Criteria for characterizing various developmental stages form another important area of research. *In vitro* techniques are being used for basic studies in micropropagation and for induction of somatic embryogenesis. In addition, clones from the program are supplied to and used by organizations to increase the effectiveness of their research in genetics, physiology, and pollution.

3.2 Denmark

Denmark has two planting zones; one covering the whole country and one special for the easternmost part. Due to the shortage of seed-orchard seeds, most of the Norway spruce now used for reforestation originate in selected seed stands. The VP program, which uses stecklings originating from clonal selections and from VM families, forms an integrated part of the overall breeding program (Wellendorf 1979, 1982, 1986). During the period 1970–1980, clones were selected as 4-year-old seedlings in commercial nurseries utilizing well-tested and adapted seed sources (Roulund 1977). The primary selection criterion was height in the nursery, and the selection intensity was about 0.001. Since 1980, most selections have been made in the breeding program's young progeny trials. Since 1988, progeny and family trials are combined with clonal tests, increasing the efficiency of the total selection procedure.

The clonal testing program, now including some 1000 candidate clones, is intended to produce well-tested clones that will also be included in the breeding population. After initial selection, cuttings are collected from the seedlings, rooted, and then transferred to a nursery where they are grown for 2 years. The stecklings from a clone are then subdivided into three groups which are transplanted to (a) an early testing area where they grow for 3 years before evaluation, (b) a juvenility archive where they are hedged in order to slow maturation, and (c) an aging archive where they are manipulated in order to hasten maturation and flowering.

A new CF program, still in its infancy, develops mixtures of increasingly selected clones on the following schedule. First, clones are selected in nurseries as described above from well-adapted seed sources. They are propagated and immediately released for production as clonal mixtures. After 10 years of clonal field tests, the number of clones in each such mixture will be reduced to between 10 and 20. Following these early tests, the best 20% of the remaining clones are selected based on their wood density. Wood density has a high heritability, and it is a key trait in Danish breeding, where the primary goal is to increase dry-matter production per unit area. The selected clones are then repropagated from the juvenility archive and established in a more extensive round of field tests at 3–4 different sites. After 12 years in those tests, the number of approved clones is only 2.5% of the original selected number. In the aging archive, ramets of the selected clones are brought into flowering, and these constitute the base for a new breeding cycle.

The main Danish VP effort, however, is VM of outstanding families, identified in progeny tests of old plus-tree selections. This has been operational for several years, and will continue to be the predominant production system for the foreseeable future. This choice is based not only on concerns about maturation of the CF clones, but it also reflects market demand.

A greenhouse seed-orchard that includes the very best tested parents from the existing breeding population has been established. The limited amounts of seeds that can be produced annually in such a SO will be vegetatively multiplied as juvenile rooted cuttings. By the year 2005, this one greenhouse SO should be able to produce all of the Norway spruce needed for planting in Denmark through VM. It is anticipated that very special genotypes will be identified and selected in the long run. Such genotypes would make it advantageous to multiline the breeding population within the 25 sublines that it presently contains.

3.2.1 Possible Drawbacks with Clones

Maturation of the clones remains a serious concern with respect to their use in reforestation. Severe maturation problems may leave VM as the only practical alternative within VP. This would reduce the genetic gains that could be achieved. Serial propagation is considered to be a biologically better way of delaying maturation than is hedging.

3.2.2 Production

Cuttings are taken in late winter and rooted during spring. They are then transplanted to an outdoor nursery, where they stay for 2–3 years before delivery. Most of the stecklings are delivered bare-root, although some containerized stecklings have been produced since 1986. In 1988, bare-root plants had 95% of the market, and they will remain predominant for the foreseeable future. Production volumes. rooting successes and unit prices are given in Tables 4, 5, and 6, respectively. Rooting success of 70% is considered commercially viable.

3.2.3 Gains

A CF gain in early height-growth of 15–20% is anticipated, based on data from 10–13-year-old field tests. For VM, the gain will be 5–10%. In addition, phenology and climatic adaptation is improved for the selected clones, and substantial improvement is expected for wood density.

3.2.4 Utilization of Clones

In Denmark, clones are deployed in three mixtures. The number of clones per mixture is not regulated; in practice, between 20 and 200 clones are included in the mixtures. Mosaic planting of monoclonal blocks is not practised, although monoclonal blocks of over 1 ha are not officially recognized as being hazardous.

The clonal mixtures are for three site indexes. They differ in average budburst time, wood density and tolerance to wind exposure. VM families are used interchangeably with seedlings, and follow the same deployment prescriptions.

3.2.5 Research

No internal research on maturation problems is under way. Some cloned genotypes are used in genetic research, and some are used by other disciplines for research in physiology and pollution.

3.3 Hilleshög

In southern and middle Sweden, five climatic (planting) zones are recognized. The Hilleshög VP program, which started in 1975, supplies these five zones with stecklings that have improved growth and whose average stem and wood quality is maintained at about that of the base population (Bentzer 1986). The H program has no direct support from an older breeding program, although some material originating from the national breeding population has been incorporated. Selections among seedlings were made in commercial nurseries, and within o-p and full-sib families supplied by The Institute for Forest Improvement.

The selection procedure and the subsequent clonal testing under nursery conditions essentially follow that of G, above. After nursery testing, where about 75% of the candidate clones are culled, the remaining clones are repropagated (a) for performance testing at four locations and (b) ramets are entered in a juvenile hedged clone archive. Two ramets per clone from the first propagation cycle are potted and are included in the breeding population. Field tests are evaluated after 4 years for height, stem quality, and various kinds of damage. About 50% of these clones are then culled, and those selected are maintained in the breeding population and repropagated from the juvenile clone archive to the cutting-donor orchard. Production of cuttings from these selected clones starts after 2 years. The field tests are further evaluated for diameter at 8 years.

The breeding population, which in 1988 consisted of about 2000 genotypes (Table 1), will eventually be reduced to about 300 genotypes based on field-test results. Selection of trees for the breeding population from clonal tests has some risks. If the proportion of additive genetic variation for the selected traits is large, then it is likely that the selected clones are appropriate for breeding purposes. If, on the other hand, mainly nonadditive genetic variation is present, then clonal selection is less efficient for breeding purposes, and the only way to fully utilize such genotypes is through VP.

Efforts are being made to speed up the maturation of clones in the breeding population, in order to start breeding work earlier. It is essential for the H program that a new population of seedlings is ready for renewed selection and testing by 1993. This is to prevent the maturation of existing clones from jeopardizing the program.

Only selection and testing of individual clones has been practised in the H program. Tested plus-trees from pedigreed crosses are not readily available, which rules out VM of tested families as an option. Hybridization between spruce species is not planned, but a number of clones have been selected from provenance-hybrid families.

3.3.1 Possible Drawbacks with Clones

From an operational point of view, the early maturation of clones constitutes the major concern. Hedging clones in the juvenile clone archive and in the cutting orchard has been the only method used for delaying maturation. For practical, fiscal, and administrative reasons, serial propagation was ruled out as an alternative. As of 1988, clones that were chronologically 17 years old were propagated from the hedged cutting orchard essentially without problems with rooting or growth habit. It is anticipated that about 2% of the clones will reach a degree of maturation annually that will make them uneconomical for further propagation. Given this modest rate of clone attrition due to maturation, bridging breeding cycles of 15–20 years with clonal propagation should be possible.

A negative genetic correlation between the ability to mature slowly and other important traits may cause problems. However, no evidence for such negative correlations has been noted to date. The slow maturation of a clone, desired for upgrading VP to CF, is antagonistic to the desire to have that same clone participate in early breeding. Selection for slow maturation may consequently cause problems in the breeding phase. A solution may be multilines, with fast-maturing lines utilizing VP or even pedigreed VM, and slow-maturing lines realizing the substantial advantages of full CF, but turning over generations less often. The costs associated with the prescribed clonal testing standards in Sweden (Hedström and Krutzsch 1982) will have a chilling effect on the development of clonal options in Norway spruce forestry.

3.3.2 Production

Cuttings are harvested from hedged donors in late summer and rooted shortly after harvest. All rooting is done in containers, and the stecklings are not removed from these until planted in the forest. Once the cuttings are in the rooting medium, all further handling is highly mechanized. After rooting, the rooted cuttings are stored during winter in the rooting house, then moved outdoors in early spring. After one growing season, the stecklings are ready for delivery. Clone identity is kept during the entire production phase, and is only removed at delivery.

Production volumes, rooting successes, and unit prices are given in Tables 4, 5, and 6, respectively. The maximum production to date was 5 million stecklings in 1984. Rooting success of 90% is considered commercially viable. The amount of labor needed during the short period of time for harvest and insertion of cuttings in the rooting medium constitutes a major organizational problem in production. Mechanization of these activities is needed. Hedge management in the cutting orchard and various administrative routines during production are other areas where improvements have to be made.

3.3.3 Gains

Judging from 5–10-year-old field-tests, an average gain in early height-growth of about 20% is expected. Compared to provenance lots of seedlings, survival of stecklings is slightly better, and adaptability and stability of the clonal mixtures is significantly better.

3.3.4 Utilization of clones

In 1988, the cutting orchard contained about 2500 clones. The most thoroughly tested of these have 400 ramets each, providing 40 000 cuttings per clone per year. Four clonal mixtures are produced. Assignment of a clone to a mixture is based on test results of the individual clones. The mixtures are assembled to meet the conditions in the five planting zones. One mixture is deployed to zones 1–2, one to zones 2–4, and one to zones 3–5. The fourth mixture is made up of clones that become dormant early in the fall, and it is deployed to sites in middle Sweden with early fall frosts. By regulation, each mixture must include a minimum of 120 clones; in practice many more are currently incorporated in each.

3.3.5 Research

A major part of Hilleshög's internal research is focused on problems associated with maturation. Donor-plant treatment to fine-tune quantity and quality of cutting production is also considered an important area. *In vitro* techniques are used to find ways to practically produce somatic embryos from mature trees. Flower stimulation techniques are being developed for the breeding population.

Clones produced in the Hilleshög program are also used for genetic studies internally, and are supplied to other organizations for yield and pollution studies.

3.4 Norway

Norway is subdivided into 11 breeding zones, and it has a VP program that is primarily focused on the more productive half of these zones. The main criteria for improvement are: climatic adaptation, quality, and volume.

In Norway, tests with rooted cuttings were initiated in the early 1950s. In the first substantial clonal test, established by Tollef Ruden in 1963, cuttings were taken from 7–8-year-old donors. Problems with poor rooting and plagiotropic growth in this test, as in a number of others, have fostered a cautious attitude toward clonal propagation. Technical difficulties with the propagation combined with difficulties of testing clones are other arguments against clonal propagation (Dietrichson and Kierulf 1982; Skröppa and Dietrichson 1986).

The VP program is at present entirely focusing on VM of progeny-tested families produced in Norway's classical tree-improvement program. Once identified, seeds of good families are produced by control pollination in existing seed-orchards, and are then further propagated by cuttings (Johnsen 1985). Note that VM will have no particular impact on the size of the present or future breeding populations.

Experiments have been initiated to develop a greenhouse seed-orchard. By combining the greenhouse SO, flower-stimulation techniques and VM, it is anticipated that this system will supply about 3 million stecklings annually by the year 2000, i.e., 5% of the national need.

Clonal tests designed exclusively to select outstanding clones are not considered at present. However, a combined family/clonal test design will be used in the near future. That will make family-index testing for forward selection in the breeding lines more efficient, and it will also provide an option for forward selection of individual clones with good precision.

Most plus trees in the breeding population have been progeny-tested by open-pollinated or polycross families. These give estimates of their general combining abilities (GCAs), but not of the performance of any specific pairwise combinations. There are no plans at present to change mating design to be able to identify favorable specific combinations, and thereby to increase the possible gain through VM of full-sib families, possibly leading to family forestry. For most traits, nonadditive genetic variation is estimated to be small in relation to additive genetic variation. Parents with good GCA are generally expected to combine well in specific pairwise combinations; consequently a mating design that would identify unusally good specific combinations is not considered worthwhile.

An expanded use of stecklings in reforestation may make large seed orchards superfluous, to be replaced by smaller breeding orchards augmented by clone archives.

3.4.1 Possible Drawbacks with Clones

Since the major thrust of Norway's spruce-improvement program, and its VP component, is not directed towards clonal forestry, maturation of individual clones is not an immediate concern. Clonal forestry is considered to have two major drawbacks, ranked in the following order: (a) the consistency and predictability of clonal performance because of maturation to untested states; and (b) high production costs. For VM, cost is a concern, but maturation may or may not be, depending on the length of time each clone donates propagules before it is replaced by a juvenile sib.

3.4.2 Production

Seedlings from controlled crosses are grown 2–3 years before the first cuttings are harvested from them. The donor plants, which are hedged, are expected to produce cuttings for at least 7–8 years. They are then replaced by new juvenile sibs or by seedlings from a better family. Cuttings are collected in late summer and rooted in the fall. All rooting is done in containers. The stecklings are ready for delivery after 1.5 years. Container plants are used by 90% of the total market today, and this percentage seems stable. Production volumes, rooting successes and unit prices are given in Tables 4, 5, and 6, respectively. Rooting success of 80% is considered commerically viable.

3.4.3 Gains

Use of VM families should improve early height by about 25% compared to ordinary seedlings and by about 15% compared to seed-orchard seedlings. These expectations are based on 8-year-old field tests. Climatic adaptation will be substantially improved as the families are tested in artificial freezing tests.

3.4.4 Utilization of Clones

At present, VM stecklings from six families are delivered to one planting zone. The plan for the near future is for each VM set to contain 10–12 full-sib families adapted to specific areas. The families can be planted in mosaics of monofamily stands or as an intimate mixture of the set of families. At present, neither clonal nor family identity is kept beyond the delivery stage.

3.4.5 Research

Since the hedging option has been adopted for the large-scale production of VM stecklings, the maturation of the family hedges and mechanization of various steps in this production process are being studied. In addition, the performance of stecklings from various donor-hedge ages is being compared to that of seedlings of the same families. Clonal material is also being used internally for genetic and physiological studies, and some clones are supplied to other organizations for studies on pollution.

3.5 Finland

Finland is subdivided into seven breeding zones for spruce, but only the southern two of these are currently included in their VP program. The primary goals for the Finnish VP program are to improve (a) growth rate, (b) climatic adaptation, and (c) quality. The work with rooted cuttings was started in the early 1960s by Lauri Kärki. Initially, it was regarded as a practically oriented research area, and in 1969 the program became more operationally oriented. It was not until 1983 that a truly separate program for mass production of stecklings was initiated. This latter program is now concentrating on VM of good families that have been selected as a part of progeny testing in the classical tree-improvement program.

Many of these good families are provenance hybrids that have performed well in southern Finland tests. The knowledge of breeding values of the parent trees is still scanty. Progeny testing has allowed some estimation of parental GCAs, but more emphasis on the performance of specific pair combinations is anticipated in the future, to be able to take fuller advantage of VM and family forestry. It is likely that combined family/clonal testing will be a standard procedure in the near future, especially for early tests under nursery conditions.

Although comprehensive clonal selections were made in the 1970s (Lepistö 1977), this approach within the VP program has been slowed. The clonal forestry that is done today focuses on CP of narrow-crown trees. Maturation seems to be a serious problem with these ideotypes, and the techniques for commercial propagation have not been satisfactorily worked out. A large-scale clonal forestry program is not regarded as realistic until maturation problems have been clarified and techniques to overcome them much improved, and the latter is believed to be unlikely before the year 2000.

It is not likely that VP will have major impact on the mainline breeding program, although the latter is under reevaluation and modification. In the proposed new program, VP is included as a separate alternative for seed-orchards. A greenhouse SO was established a few years ago. Successful flower induction of young grafts has been accomplished, and VM of full-sib families is under way. However, this greenhouse SO is regarded more as a clone archive than as an elite SO.

3.5.1 Possible Drawbacks with Clones

Problems with maturation are the major concern. Hedging of donor plants has been tried, but maturation has not been sufficiently delayed by hedging to support a commitment to operational production. It now appears that serial propagation is more suitable for maintaining juvenility of donor plants than is hedging. The difficulty of testing clones and the present lack of proof of genetic gains by cloning are other points of concern leading to caution. Even for VM, where maturation is not a concern, the high production costs compared to seedlings make the acceptance of stecklings difficult.

3.5.2 Production

Cuttings are now harvested during winter from serially propagated donor plants grown in plastic greenhouses; these technique advances have improved the quality of the cuttings and made harvesting more efficient. All rooting is in containers during spring, and the stecklings are delivered 2 years later. In Finland, container plants are used by about 45% of the total market. Production volumes, rooting successes, and unit prices are given in Tables 4, 5, and 6, respectively. Rooting success of 80% is considered commerically viable. There are still many technical problems associated with steckling production that need to be solved.

3.5.3 Gains

A 20–30% increase in volume growth is estimated for VM, and better survival and improved quality, the latter particularly for narrow-crown trees, is expected.

3.5.4 Utilization of Clones

VM families are delivered as mixtures to those planting zones where the families have been tested and selected. Currently, stecklings are intimately mixed with seedlings during planting. In the near future, and particularly if family forestry is adopted, mosaics of steckling and seedling blocks will probably be planted. However, no official strategy for the use of either pedigreed or VM stecklings has been worked out in Finland.

3.5.5 Research

At present, very little research is conducted on the utilization of spruce clones in Finland. Some studies of donor-plant management for maintaining juvenility have been initiated, and *in vitro* research focusing on micropropagation is in progress.

4 Comparison of the Programs

For various reasons, different emphases have developed in the different programs. Common to them all, however, is the conviction that clonal forestry has a greater potential than vegetative multiplication of untested clones (VM) to capture gains in essentially all traits under selection. Equally unanimous is the concern about maturation of clones, and the problems that this presents for full clonal forestry. At present, the D, N, and F programs have concluded that maturation prevents CF from being an operational alternative. On the other hand, the G and H programs have experience that indicates that maturation, although serious, can be sufficiently controlled for clonal forestry to be the preferred alternative. German legislation against VM and the lack of sufficient progeny-tested families in the H program are factors contributing to the pursuit of CF in these programs.

Research devoted to studying and eventually solving the maturation problem is surprisingly meager. The G and H programs are, for obvious reasons, those that most emphasize and perform such research.

Unique features of the G program are the continued selection of new clones from commercial nurseries, and the great effort put into species hybridization. Repeated selections from the same populations are unlikely to add much to genetic gain, particularly if these are mostly serving the (perhaps important) function of replacing maturing clones with juvenile ones. The introduction of first- and later-generation hybrids could substantially increase the available genetic variation.

The H program is the only one to have established a substantial clonal hedge orchard (25 ha) where the appropriate manipulation of donor plants is a key factor in delaying maturation. Most new selections for this program will originate from families produced by crosses between clones with excellent performance records. High priority is given to research on flower induction in these clones, many of which allocate resources to wood rather than to flowering in field tests.

Of the three programs concentrating on VM, the N program has by far the largest number of tested parent trees from which families can be produced (Table 2), and thus the N program has the largest selection differentials available for family forestry. Other significant features of the N program are its emphasis on selection for adaptability and the maintenance of seedling families as hedges for cutting production. If the performance of stecklings from these hedges remains good as the hedges become chronologically older, it will be relatively easy for the N program to shift from VM to CF.

In the N and D programs, options for CF are also kept open by the combined family/clonal testing procedures that are being introduced. Then, if maturation can be mastered, tested clones will be available for rapid production and deployment. Although a combined family/clonal test may not be as efficient for clone evaluation as a pure clonal test, it will certainly constitute a good compromise between flexibility, efficiency, and cost.

A feature of the D program is the expressed goal of maximizing dry-matter production. This introduces basic wood density as a key selection criterion, which may have a major impact on overall breeding strategy. The Danish program is also the one where VP has been most integrated with breeding strategy, keeping many options open in the future. In the D program, cuttings are produced from families of hedged seedling ortets, not because this is a method superior to serial propagation for maintaining juvenility, but because it is recognized as much more efficient to administer.

In Finland, technical problems with VP, and particularly with long-term clonal propagation, have made them hesitant as to which way to commit themselves. Growing donor plants in plastic houses has improved the quality of the stecklings. The focus of VP on narrow-crown ideotypes of Norway spruce is a unique feature of the F program. Also, VM families are mixed with seedlings during planting. This is not primarily for reducing potential risks of failure by the stecklings, but rather to enlarge the area on which stecklings can be planted, given their low numbers, high costs and likely values.

5 Operational Production

VP of spruce is still in its infancy. Although the proportion of stecklings to seedlings is about 10% in Germany and Denmark at present, smaller percentages are used in the whole of the Norway spruce region. This contrasts to radiata pine (APM, Australia), redwood (Rydelius and Libby, Chap. 9, Vol. 2), and eucalypt (Zobel, Chap. 7, Vol. 2) programs, all of which are or soon will be 100% clonal. To produce substantial quantities of spruce stecklings, e.g., more than 10 million annually, not only biological problems have to be under reasonable control, but also financing and production efficiencies will be major concerns. With the production systems used for spruce today, which are mainly based on manual handling, only marginal economies of scale will be obtainable.

Mechanized equipment for various stages of production will need to be developed, along with improvement in biological techniques. Rooting success is of paramount importance from an economical point of view, and each 1% of rooting failure will increasingly hamper the future use of stecklings.

Technical knowledge of rooting Norway spruce cuttings has greatly improved during the past 20 years, and it is likely to continue to do so. With further expanded knowledge of the genetics of clones, of risks, and of utilization, it is logical that spruce clonal forestry will substantially expand and that it will actually capture most of the gains that are theoretically possible.

Acknowledgments. Discussions with J. Kleinschmit in Germany, H. Roulund in Denmark, T. Skröppa and Ö. Johnsen in Norway, J. Mikola and M. Lepistö in Finland are highly appreciated.

References

Bentzer B (1986) Operational cutting production of Norway spruce *Picea abies* (L.) Karst Mitt Bundesforschungsanst Forst Holzwirtsch (Hamburg) 154:134–141

Dietrichson J (1979) Norway spruce provenanace trials in Nordic countries. In: Proc IUFRO Joint Meet Working Parties Norway spruce provenances and Norway spruce breeding, Bucharest, Rom Lower Sax For Res Inst, Dep For Tree Breed, D-3513 Staufenberg-Escherode, FRG, pp 3–9

Dietrichson J, Kierulf C (1982) Selection of eight-year-old Norway spruce [*Picea abies* (L.) Karst.] plants in progeny trial and mass production by cuttings. Medd Norsk Inst Skogforsk 38 (1):1–28

Fowler DP (1979) Norway spruce provenance experiments in North America. In: Proc IUFRO Joint Meet Working Parties Norway spruce provenances and Norway spruce breeding, Bucharest, Rom Lower Sax For Res Inst, Dep For Breed, D-3513 Staufenberg-Escherode, FRG, pp 29–35

Hedström BS, Krutzsch P (1982) Regulations on clonal forestry with *Picea abies*. In: Proc IUFRO Joint Meet Working Parties genetics about breeding strategies including multiclonal varieties, Sensenstein, Germany. Lower Sax For Res Inst, Dep For Tree Breed, D-3513 Staufenberg-Escherode, FRG, pp 109–112

Johnsen Ö (1985) Successive bulk propagation of juvenile plants from full-sib families of Norway spruce. For Ecol Manag 11:271–282

Kleinschmit J, Müller W, Schmit J, Racz J (1973) Entwicklung der Stecklingsvermehrung von Fichte (*Picea abies* Karst.) zur Praxisreife. Silv Genet 22:4–15

Lepistö M (1977) Vegetative propagation by cuttings of *Picea abies* in Finland. In: Vegetative propagation of forest trees – physiology and practice. Inst For Improv & Dep For Genet, Coll For, Swed Univ Agric Sci, pp 87–95

Roulund H (1977) Vegetative propagation of forest trees at the arboretum in Hörsholm, Denmark. In: Vegetative propagation of forest trees – physiology and practice. Inst For Improv & Dep For Genet, Coll For, Swed. Univ Agric Sci, pp 103–128

Samuelson KR (1979) Possibilities of seed orchards in Norway spruce. In: Proc IUFRO Joint Meet Working Parties Norway spruce provenances and Norway spruce breeding, Bucharest, Rom Lower Sax Res Inst, Dep For Tree Breed, D-3513 Staufenberg-Escherode, FRG, pp 129–138

Skröppa T, Dietrichson J (1986) Genetic variation and ortet/ramet relationships in a clonal test with *Picea abies*. Scand J For Res 1:323–332

Wellendorf H (1979) How can the present knowledge of breeding be used for optimization of spruce breeding programs. In: Proc IUFRO Joint Meet Working Parties Norway spruce provenances and Norway spruce breeding, Bucharest, Rom Lower Sax For Res Inst, Dep For Tree Breed, D-3513 Staufenberg-Escherode, FRG, pp 256–279

Wellendorf H (1982) Cutting propagation as a tool for within-family selection of sub-lines of Norway spruce. In: Proc IUFRO Joint Meet Working Parties Genetics about breeding strategies including multiclonal varieties, Sensenstein, Germany. Lower Sax For Res Inst, Dep For Tree Breed, D-3513 Staufenberg-Escherode, FRG, p 30

Wellendorf H (1986) Danish spruce breeding. In: Proc IUFRO Joint Meet Working Parties Breeding theory progeny testing seed orchards, Williamsburg, Virg. NC State Univ Ind Coop Tree Improv Program, Raleigh, NC, pp 239–249

Werner M (1977) Vegetative propagation by cuttings of *Picea abies* in Sweden. In: Vegetative propagation of forest trees – physiology and practice. Inst For Improv & Dep For Genet, Swed Univ Agric Sci, pp 97–102

Werner M, Pettersson H (1981) Klonskogsbruk med gran. In: Föreningen skogsträdsförädling och Inst Skogsförbättr, Årsbok 1981. Inst Skogsförbättr, Box 7007, 75008 Uppsala, pp 118–153

Zobel B, Talbert J (1984) Applied forest tree improvement. John Wiley & Sons, New York, 505 pp

7 Clonal Forestry in the Eucalypts

B.J. Zobel[1]

1 General

The earliest operational use of rooted cuttings in plantations of forest trees was not with the eucalypts, since rooted cuttings were used over 100 years ago in *Cryptomeria japonica* (Ono 1882). This kind of regeneration has been common in plantations of *Populus* spp. and *Salix* spp. Currently, however, the largest operational clonal forestry programs are with several species in the genus *Eucalyptus*. For nearly 20 years, large plantings of rooted cuttings of eucalypts have been made in the tropical and sub-tropical areas of Africa (Delwaulle 1985), South America (Campinhos and Ikemori 1977), and in other areas such as the Mediterranean region and South Africa. A few programs are very large – for example, one company in Brazil plants nearly 30 million stecklings per year and its program is expanding annually. The largest programs have received a great deal of publicity; an example is Aracruz Florestal in Brazil, which has achieved breakthrough successes with rooted cuttings (Fig. 1); its foresters have received the prestigious Wallenberg Prize from Sweden in recognition of their pioneering efforts (Wallenberg Foundation 1984).

The shift from the use of seedlings to clonal *Eucalyptus* forestry did not happen rapidly. A few far-sighted and imaginative people tried rooted cuttings. Success was marginal, yet good enough to invite more trials. There were so many methodological problems, such as what kind of cutting to root, the best rooting medium, moisture distribution while rooting, shade, required time to plant, and others, that many unsuccessfully tried rooting cuttings and became discouraged. A few persisted, and once the methodology became operational, they switched from seedlings to stecklings very rapidly. This switch is not general, and still today many foresters are suspicious of this "new method" and are reluctant to try clonal forestry. Many have been discouraged because they found that each organization has to develop methodology to fit its own unique situation and there is no standard universal best method. Because of recent publicity about success achieved from both the Congo and Brazil, many organizations are trying to root cuttings of eucalypts, even with species like *E. globulus*, which is reported to be a poor rooter. Clonal forestry in the eucalypts is now well recognized, but sometimes is not used as widely as it should be.

[1]College of Forest Resources, North Carolina State University and President, Zobel Forestry Associates, P.O. Box 37398, Raleigh, NC 27627, USA

Clonal Forestry II, Conservation and Application
Ed. by M.R. Ahuja and W.J. Libby

Fig. 1. A *Eucalypt* plantation derived from 4-year-old cuttings in Aracruz, Brazil

The planting of stecklings is spreading very rapidly among organizations using certain species of eucalypts, as well as among some using other species of conifers and hardwoods. The value of clonal forestry has been fully explained elsewhere in this book but it is essential to re-emphasize here that vegetative propagation is *NOT* a breeding method to develop better genotypes. It is *ONLY* a method to mass-produce desired trees and to get the increased benefits by using outstanding and known good genotypes. It does not create anything new or directly produce improvements in advanced generations. It does enable taking advantage of differing genes through maturation and using the knowledge gained from clonal testing, which indicates the value of the genotype when it is used as a clone.

2 Some Misconceptions

There is no doubting that the use of vegetative propagation in some of the eucalypts is now a major and integral part of silviculture and forest management. But there is a belief by many persons that all eucalypts will produce sprouts, which yield juvenile shoots that will then root easily. This is not true at all. For example, some species (examples are *E. regnans* and *E. nitens*) do not sprout readily or at all. Some species (*E. camaldulensis* and *E. deglupta*) sprout vigorously and root easily, while other species sprout readily but root with difficulty (*E. globulus*). No general statement can be made about the utility of rooted cuttings in the genus *Eucalyptus* because of the large variability among species (or even within species) but it is becoming the standard method of operational propagation in species like *E. grandis*.

Another misconception is that there is one methodology of rooting that works for all *Eucalyptus* species in all areas. The facts are that each organization must work out techniques suitable for clonal production for its own conditions. This is not easy; causes for differences between success and failure are often subtle and interrelated. Some circumstances are downright puzzling. One company in South America tried the "standard method" and had good success (75% rooting) the first 2 years. The 3rd year rooting success dropped to 5%. A whole group of experts tried to determine the problem, but in the 4th year there was no rooting at all. The nursery was moved some distance away, and the rooting percentage is again 80%. We suspected some type of water contamination at the old location and made every test possible, including trucking in water from elsewhere. No adverse agents were found on which to blame the sudden failure of rooting the 3rd and 4th years. The point being illustrated is that vegetative propagation (rooting) of the eucalypts is not always easy, and that one cannot just blindly follow the methods that are successful for another organization.

Many people think that all eucalypts have similar and good wood. However, variation in the wood among the many hundreds of species of *Eucalyptus* is huge, some having wood with a density nearly as low as balsa and others with a specific gravity over one, so that their wood sinks when put in water. Many times I have asked the head of a new eucalypt program "What sort of wood will be produced?" and received the answer that the species has good wood in its indigenous range and therefore it will be good in his plantations. Many foresters are unaware that wood properties can vary greatly, depending not only on the tree's genotype, but also upon the environment in which the tree is grown. We have observed tests of the wood of only one or two trees from a species and, on the basis of the results, accept or reject the species as a whole. The chance of sampling a tree with nonaverage wood is very high, which then results in an incorrect decision. For example, the wood specific gravity of individual trees of *E. grandis* in a plantation has been observed to vary from 0.35 to 0.68. If one of the individuals near the extremes is chosen, a nonsense decision will result. Most species of eucalypts have a large between-tree variation. As will be discussed later, one major benefit of vegetative propagation that can be easily achieved is to have more uniformity in the wood produced and clonal selection for desired wood can be effectively done.

It is quite remarkable how many persons are convinced that a good-looking tree will automatically produce good clonal copies when grown in a plantation. It is common to search hard for a good eucalypt phenotype and then to use it as a donor in a clonal planting program with no further testing. This is a mistake indeed! While it is true that most good phenotypes will produce good trees from rooted cuttings, there is a significant percentage that will not. Because of this, *all clonal rooted cuttings must be field-tested before they are used in operational plantings.* It is mandatory for all companies with which I work to test *each clone* on *each* (significantly) *different site* before it is used in a large operational clonal program. Responses of genotypes to cloning and genotype × environment interaction differences among greatly differing sites are so large that full clonal success will never be achieved in the eucalypts if these are not taken into consideration.

Two examples illustrate this need: one of the best phenotypes produced a very superior clonal planting for the first 1.5 years of its test. A lot of wind and rain then occurred and many of the trees (about 9 m tall at 18 months) fell over. When we obtained a tractor and pulled some of the trees up, we found that every member of this particular clone had only two major roots that were approximately 180° apart. In the wet soil the wind just pivoted the trees out of the ground. Without testing long enough, operational planting of this clone would have been a disaster. In another clone, every tree forked at about 6 m height. The two forks grew equally well. However, after a heavy wind, all of the forked tops broke apart, reducing the planting to a field of 6-m poles. The original phenotype from which the cuttings were made appeared to be one of the very best. If that clone had been used operationally without proper testing, serious losses would have occurred.

3 Some Advantages from Rooting Eucalypts

In addition to greater uniformity, the use of clonal propagation of *Eucalyptus* has shown major advantages resulting from a full capture of the genetic potential of desired genotypes.

One of the greatest benefits from clonal propagation is in volume gain since growth rate has a relatively large amount of nonadditive variance, which can best be captured through vegetative propagation. One organization in Brazil has increased the average volume production from 33 m^3/ha/yr when unimproved seedlings were used to about 70 m^3/ha/yr with cuttings from selected trees. This equates to a weight from 7.9 tons of pulp/ha/yr to 18.5 tons from the rooted cuttings (Brandao 1984). The greater percentage increase in tons of pulp relative to volume is due to use of the desired specific gravity and the capturing of greater cellulose yields possible in clonal forestry. Individual clones have been found in tests where ideal management was used that produced over 100 m^3/ha/yr on only moderately good sites (Brandao 1984). Such growth is not thus far obtainable from seedlings.

The second and third most important advantages are in the uniformity and desirability of the wood produced. Clones with the most desired wood will be planted together, making a homogenous wood furnish that is a delight to the mill people. With intensive effort, Aracruz Florestal has developed "a group of clones with very uniform wood, which results in every batch of pulp being quite like every other. By tailor-making the wood density through clonal choice to the desires of the customer and by capturing the differences in cellulose yield (which can best be modified through vegetative propagation) Aracruz has been able to increase the amount of desirable cellulose per m^3 of wood from its improved stock by 25% over the variable wood produced by unimproved material . . . ." (Zobel 1988). Cellulose yield in pines is primarily influenced by nonadditive components of genetic variation; only a small fraction ($< 5\%$) can be captured through a standard seed orchard, but most of it can be used through vegetative propagation (Zobel et al. 1966; Jett et al. 1977). I assume the same holds for the eucalypts. In the South American eucalypt programs, clones are nearly always chosen to have desired wood along with superior growth, form, adaptability, and disease resistance. Different final products are best produced from specific wood properties. This can be easily achieved with the use of rooted cuttings. The most advanced example of this potential was reported by the Wallenberg Foundation (1984). A sample of the results obtained are summarized from one of their tables. All data reported are for 7-year-old *E. grandis* and hybrid trees ready for harvest; both bark and wood are included except as indicated by an asterisk.

Item	Initial *seedling forest*	Improved rooted *cutting forest*	Improvement *Amount*	*Percent*
Yield (m^3/ha/yr)	33	70	+ 37	+ 112
Productivity (tons of pulp/ha/yr)	7.85	18.45	+ 10.60	+ 135
Basic density (kg/m^3)	460	575*	+ 115	+ 25
Pulp yield (percent)	48	51	+ 3	+ 6
Pulp content (kg pulp/m^3 wood)	238	293*	+ 55	+ 23
Wood consumption (m^3 of wood/ton of pulp)	4.20	3.41*	− 0.79	− 19

*Wood only, bark not included.

The improvement in wood uniformity and desired quality following use of better genetic stock and rooted cuttings was striking in 1984 but is even considerably better now in 1992.

To repeat, the key characteristic desired in wood is uniformity. For example, in 1984 the range in wood density of individual trees of *E. grandis* grown from seed varied from 300 to 900 kg/m^3. The clonal material planted varied only from 500 to 600 kg/m^3 (Wallenberg Foundation 1984). Currently, the average has been shifted downward some to meet market demands, and the range among clones is even better than before, being less than reported in 1984.

The use of rooted cuttings can be very effective in the reduction of pests. The *Eucalyptus* canker disease *Cryphonectria cubensis* (formerly *Diaporthe cubensis*) is severe in some areas and for some species and provenances. It became so prevalent in parts of Brazil that reducing the use of *Eucalyptus* was being considered, and there was a reluctance on the part of investors to establish new

Eucalyptus forestry operations there. The plantings of companies in coastal Brazil had infections of canker ranging from 5 to 85%. After suitable testing, resistant clones were deployed; now many thousands of hectares can be found without a single infected tree. Control of other pests on the eucalypts, such as some leaf-eating insects, is successful by similar deployment of tested clones.

One advantage of clonal forestry, sometimes the best of all, is the opportunity to match clones with site and silvicultural treatment. This is especially true for eucalypt clones, which generally show a strong genotype × environment interaction when environments vary considerably. Clonal forestry in the eucalypts is tree farming at its best. Not only are clones matched to site, but some of the companies in Brazil, Colombia, and Venezuela are matching clone to kind of fertilization and even to the timing of fertilization. Clones respond differently to subsoiling and competition control. For example, some have dense crowns that shade the competition quickly; these are planted on sites with heavy undergrowth, thereby reducing the number of cleanings necessary to grow the crop.

In some genera, rooted cuttings have root systems that are qualitatively different from those formed by seedlings. For example, the roots of *Bombacopsis quinata* cuttings are shallow and fibrous, without formation of the massive tap root generally formed by seedlings. This limits the use of rooted cuttings to the moister sites; they are drought prone on the drier sites. Generally, however, the roots of eucalypt cuttings develop similarly to those of seedlings. We have had little complaint about the root form from clonal plantings in *Eucalyptus*.

When breeding, especially when hybrid parents are involved, occasional trees with unique and rare gene combinations are formed for a favorable expression of important traits. Some of these probably result from a nonadditive combination of traits. Some are greatly superior and different from either parent and these kinds of gene combinations normally recombine in less favorable associations, in both sibs and offspring of the selected clone. It may take many generations of line breeding to obtain stable sexual offspring that give such a favorable combination as can be found and used clonally from the highly variable sexual propagules now used. Therefore, the chance genotypes, more common when hybrids are crossed than when pure species are used, cannot be exploited from a seed regeneration program but they are ideal for a clonal program. Some of our very best eucalypt clones arose as chance, "non-repeatable" gene combinations from crosses or backcrosses among hybrids.

4 Some Disadvantages of Using Clones

Although clonal forestry in the eucalypts has many benefits, there are some distinct problems. None of these is insurmountable, but they must be recognized and adjusted for if eucalypt clonal forestry is to be totally successful. Generally, the emphasis has been on the advantages of clonal forestry and its problems have been minimized. Some of the problems are real and biological, while some are merely the results of assumptions and misunderstandings by foresters and the general public. Nevertheless, the latter are important and can have a major dampening effect on clonal forestry in the eucalypts if they are not addressed.

The major biological problem is the "fallout" of genotypes that occurs in satisfying all the strict criteria for operational clonal planting. A large number of selected trees must be initially available if one is to end with the desired number of clones suitable for operational planting. In Colombia and Venezuela, several thousand initial selections were made to operate programs where 50 *E. grandis* clones were ultimately used. Aracruz Florestal in Brazil had 100 best clones as its objective. To obtain these, over 5000 selections were initially made in plantations. In addition, 1254 provenances and families have been tested (Brandao 1984). Once a tree is good enough to be selected, it then has to: (1) sprout well (2) root well, above 70% (3) grow well when tested clonally, (4) have good tree form, (5) have desirable wood, (6) have adaptability to at least some of the particular sites being considered for planting. Many otherwise superior genotypes are lost because the tree does not fit just one of the criteria listed. Our breeding program maintains the "rejected" superior genotypes, and through sexual recombination, will produce additional trees, some of which will be suitable clones. If this is not done, erosion of the available usable genotypes may occur, and long-term improvement will be greatly hindered.

There is a strong antagonism against *Eucalyptus* plantations by many environmentally oriented people. Chapter 10 in *Growing Exotic Forests* (Zobel et al. 1987) deals specifically with some of the supposed problems. Most of the objections-forming deserts, poisoning soil, causing erosion etc. – are not biologically sound. However, the major objection by the public to monocultural planting is real and must be dealt with. It can be handled biologically, if proper care is taken right from the start of the program. (Phases of this are covered in the next section of this chapter). If laws with severe restrictions against clonal usage should be passed in tropical countries similar to those in Germany and some other countries (Muhs 1982), operational planting of rooted cuttings of *Eucalyptus* would cease and the benefits of clonal forestry would be lost. It is absolutely essential that controls related to large-scale planting of clones be based on fact, and that they be realistic.

Although previously listed as an advantage when used properly, the strong genotype × environment interaction of *Eucalyptus* clones may create problems. Some eucalypts are very sensitive to environmental variations. Therefore, a strong clonal field testing is mandatory to match clones with sites; results are particularly evident when the usual clonal block planting is done (see below). It is essential that any important genotype × environment interactions be taken into account when block planting is done. When using seedlings from improved families, the testing and use of significant G × E interaction of individual trees is not possible. Once known for a given clone, G × E interaction is a major determinant as to where and how that clone will be used; this is especially so in the eucalypts. One rarely finds universally good eucalypt clones suited to significantly different environmental conditions and sites. This means that a successful operational vegetative-propagation operation in the eucalypts must be accompanied by intensive developmental and testing programs. Many organizations are either unwilling or unable to do this; as a result their clonal programs are inferior.

5 The Future of Clonal Propagation in the Eucalypts

There is no question that vegetative propagation will be the major method of regeneration in many *Eucalyptus* species. Many of the major companies planting eucalypts in the tropical and subtropical areas are now using clonal forestry or moving in that direction; this may soon include the huge planting programs for charcoal production in areas such as Minas Gerais, Brazil, where nearly 100 000 ha are planted each year. Unless legislation is passed to interfere, the yields from *Eucalyptus* forestry will continue to escalate as more greatly improved clones are developed, are matched to silvicultural methods, and are grown on sites where they are well adapted. None of our clients, which include many of the larger eucalypt producers in the world, is considering a return to the use of seedlings.

The major unknown is how rapidly the organizations growing *Eucalyptus* realize the need for an intensive breeding and developmental program from which future improved clones will be secured. There is currently some tendency to do for the eucalypts as was done for poplars in the early years; i.e., to be satisfied with only developing and using clones from the first generation of improvement. Members of a clone are no better than the tree from which the cuttings are obtained. Therefore, a clonal program must always be accompanied by a developmental (research) program if it is to produce better genotypes for additional gains over those obtained from the initial selections.

If one catalogued all the breeding programs in *Eucalyptus* where clonal planting is used, it would vary all the way from none through various classical approaches to an intensive breeding program using sublining. An example of the latter is Aracruz' long-term breeding program, now well developed. This consists of sublining *E. grandis* and *E. urophylla*. Selections have been made to establish 20 sublines of 20 selections each, or a total of 400 intensive, (mostly) second-generation selections per species. At the 4th generation, the best trees from the sublines of each species will be crossed, giving "super-hybrids", from which additional selections will be made for clonal testing. In addition, crosses are being made between promising hybrids in the eucalypts. Since only individual outstanding trees are tested and then used in a clonal program (family performance is not so important), any outstanding genotypes from the hybrid crosses can be used as clones for the production program.

Specific breeding programs used in the eucalypts with clonal programs are not presented in this chapter. Suffice it, instead, to reiterate and emphasize: a *clonal program in Eucalyptus cannot be successful without an accompanying intensive breeding program.* Remember, unless there are advantages associated with the advanced maturation state, the rooted cuttings are no better genetically than the trees from which they come.

Decisions must be made primarily on intimately mixed vs. clonal-block plantings, on the numbers of clones necessary for reasonable safety from monocultural disasters, and clonal block size when blocks are used. Although no research results can be cited to answer these questions, my long and widespread experience with growing clonal *Eucalyptus* has led to the following conclusions (Zobel et al. 1987; Zobel 1992). At our current level of knowledge, mixed clonal plantings are not suitable for production forestry. In addition to the difficulty of making the mixtures for a large planting program, clones have different growth curves. We find in Brazil, for example, that about 20% of the clones planted in mixtures never mature to harvest because they are slow starters that will be suppressed by the faster starters, even though the slow starters will be as large or larger at time of harvest (6 years) when grown in pure clonal blocks. In the eucalypts, with such short rotations, survival near 100% is usually obtained and any significantly less stocking will result in reduced economic benefits. To accept 80% survival just to mix clones is foolish indeed. Also, if some severe environmental malady or pest attacks one or two clones in the mixture, the trees of that clone cannot be effectively salvaged by thinning and replanting which cannot be done. However, if grown in blocks, the clonal block that is damaged can be harvested early and replanted. This means that the blocks must be of an operable size for silvicultural manipulation. We use 10 to 20 ha as suitable block size.

An often-debated question relates to how many clones to use. It is, of course, economically best to use the fewest number that is safe because gain is dependent on the intensity of the selection differential. But no one can define what is "a safe number". Most figures published relative to clonal numbers to be used are based upon emotion or political considerations. For example, planting 500 clones in mixture, cited by Muhs (1982), is patently ridiculous and there is no biological basis for requiring this large number of clones. If one were forced to follow this law, and similar ones in several other countries, clonal forestry in the eucalypts would not be used, and one could obtain better returns by using seedlings. Oppositely, the use of one or two of the best clones, as we have seen proposed, is too dangerous to seriously consider. The decision as to the number of clones that is safe enough, yet give the best gain, can best be made when based upon an intimate knowledge of the species and of what the likely risks of a dangerous monoculture will be. As a general rule, based upon how much risk can be taken to obtain a given amount of gain. I recommend that at least 15 unrelated well-adapted (based upon tests) clones are sufficient in *E. grandis* and *E. urophylla* if planted in blocks that are intermixed. Each organization must assess its attitude as to risk vs. gain, but if a very high number of clones are used without experience just because a large number sounds good politically, the potential for outstanding performance (gain) will be greatly decreased.

If a proper assessment of gain vs. risk is not made, clonal forest planting will fall short of the benefits of which it is capable.

References

Brandao LG (1984) The new eucalypt forest. In: Proc 1st, Symp Marcus Wallenberg Found, Falun, Swed, pp 3–15

Campinhos E, Ikemori YK (1977) Tree improvement program of *Eucalyptus* spp.: preliminary results. In: 3rd World Consul for Tree Breed, Canberra, Aust, pp 717–738

Delwaulle JC (1985) Clonal plantations of hybrid *Eucalyptus* in the Congo. Bios For Trop 208(2):37–42

Jett JB, Weir RJ, Barker JA (1977) The inheritance of cellulose in loblolly pine. In: TAPPI For Biol Comm (its committee) Meet, Madison, WI, 4 pp

Muhs JH (1982) Identifying problems and legal aspects of multiclonal mixtures. In: IUFRO Joint Meet Work Parties Genet, Escherode, FRG, 17 pp

Ono M (1882) Which grow faster, seedlings or cuttings in the case of *Cryptomeria* or hinoki cypress? Abst Jpn Lit For Genet (ed) Related Fields 1, 1:22

Wallenberg Foundation (ed) (1984) The new eucalypt forest. In: Proc 1st, Symp Wallenberg Found, Falun, Swed, 64 pp

Zobel BJ (1988) *Eucalyptus* in the forest industry. Tappi 71(12):42–46

Zobel BJ (1992) Vegetative propagation in production forestry. Forestry 90(4):29–33

Zobel BJ, Stonecypher R, Browne C, Kellison RC (1966) Variation and inheritance of cellulose in the southern pines. Tappi 49(9):383–387

Zobel BJ, van Wyk G, Stahl P (1987) Growing exotic forests. John Wiley & Sons, New York, 508 pp

8 Development of Plantlings and Stecklings of Radiata Pine

J.A. Gleed[1]

1 Introduction

Tasman Forestry Limited (TFL) manages 200000 hectares of *Pinus radiata* (D.Don) forests in New Zealand primarily for clearwood production, which necessitates intensive pruning to 6.0 m of the final crop stems. Pulpwood and framing timber come from intermediate thinnings and unpruned top logs.

Silvicultural regimes are supported by a dynamic tree-improvement programme producing both open- and controlled-pollinated seed from four seed-orchards. The overall strategy of the tree-improvement program is "to plant the best available combination of genes on the best sites at the earliest times". To this end a variety of plant propagation systems are used at TFL forest nursery to produce 6–7 million plants per year for forest establishment (over 7000 hectares). All annual planting stock is sourced directly or indirectly from the seed-orchards (Fig. 1). Table 1 summarizes past and anticipated production levels of the various stock types.

The continuous improvement of TFLs genetic resource constantly challenges the perceived advantages for various planting stock types. The advantages for adolescent stecklings as reported by Arnold and Gleed (1985) had by 1990 been diluted by the greater availability of advanced second-generation seedlings and increasing amounts of cloned tissue-cultured plantlets. Adolescent stecklings are likely to become "in favor" again when they can be sourced from within "tested" clonal stands.

The vegetative multiplication of control-pollinated progenies by juvenile stecklings from nursery stool beds is likewise destined to diminish as large quantities of control-pollinated seed become routinely available (Arnold 1990).

At present all four stock types are complementary with production methods under constant development and review with the major thrust toward full clonal forestry via tissue-culture production.

[1] Tasman Forestry Limited, Private Bag 3031, Rotorua 3200, Bay of Plenty, New Zealand

Clonal Forestry II, Conservation and Application
Ed. by M.R. Ahuja and W.J. Libby

Table 1. Approximate annual production levels 1985–1994 (millions)

	Seedlings		Stecklings		Tissue culture
	OP	Control-pollinated (family)	Adolescent (family)	Juvenile (some clonal)	Clonal
1985/89	5.0	0.1	1.0	0.5	0.3
1990	4.0	0.1	-	0.5	0.5
1991	4.0	0.1	-	1.0	1.0
1992	3.0	0.1	0.5	1.0	[a]2.0
1993	3.0	0.2	0.5	1.0	[a]2.5
1994	3.0	0.2	0.5	0.5	[a]3.0

[a]Increasing quantities of tested clones.

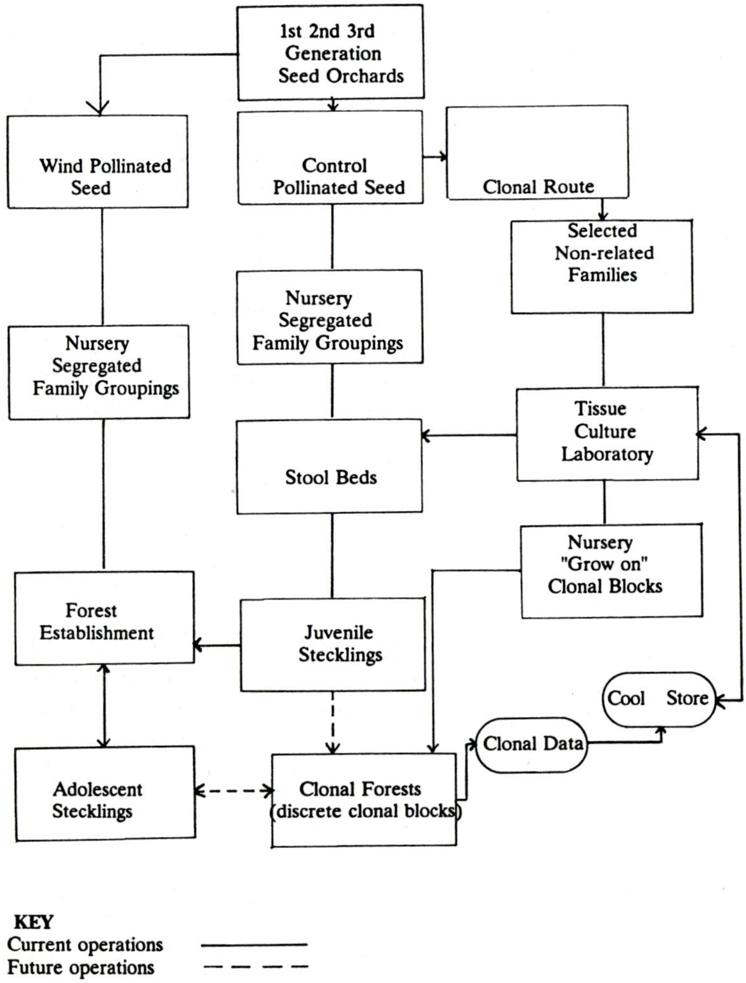

Fig. 1. Plant stock production pathway (Tasman Forestry Limited)

2 Vegetative Propagation (Radiata Pine)

At TFL vegetative propagules are sourced from four distinct origins viz:

1. Adolescent stecklings – derived from cuttings taken from selected genotypes within genetically improved plantations. Age of ortets (from seed) can range between 3–7 years with the emphasis on the 4–5 years. Selection criteria is based upon vigor, form, branching, and health.
2. Juvenile stecklings – derived from "hedged" seedlings of full-sib families, which are normally in production for 3 years, at which time they are discarded in favor of improved genetic material. Hedges may also be derived from tissue-culture plantlets.
3. Tissue-culture plantlets (embryo) – derived from embryos from nonrelated control-pollinated families within progeny-tested seed-orchard clones. Cloned shoot material is mass propagated and rooted under laboratory sterile conditions, grown on in the greenhouse, and lined out in the nursery for development into plantable stock.
4. Tissue-culture plantlets (closed buds) – derived from buds taken from 3–6-year-old genotypes as for adolescent stecklings. Selection ratio is higher with selected genotypes frequently sourced within progeny trial areas.

2.1 Adolescent Stecklings

Radiata pine plantations when grown from adolescent stecklings exhibit superior traits including improved straightness, less taper, thinner and less branches. This is especially so on fertile sites, which tend to exaggerate the gross adolescent form of the species. These improved traits readily translate to economic advantages for the forest grower of pruned logs (Arnold and Gleed 1985).

These advantages are achieved, despite the fact that stecklings may cost up to five times the cost of seedlings because fewer stecklings (750 s/ha) are planted than seedlings (1200 s/ha) and all subsequent silvicultural operations are more efficient due to lower work content. In one study at TFL (Arnold and Gleed 1985) the costs of radiata pine plantations grown on clearwood regimes established with seed-orchard seedlings or adolescent stecklings was compared. Over the first 7 years following establishment, by which time all thinning and pruning was completed, the discounted savings were $NZ 200 in favor of stecklings. The savings included ($NZ per hectare): planting $ 56; weed control $ 62; waste thinning $ 87; pruning $ 208. Adolescent stecklings cost an extra $ 120.

When such savings are carried on to a 7000 hectare per year forest establishment and management program (TFL average) the above operational savings would amount to more than $NZ 1.0 million per year. It is this magnitude of cost savings that TFL aims to capture through its vegetative propagation and tissue-culture programs. Clearwood regimes require a large financial commitment to pruning, and thinner branches contribute significantly to cost savings, the extent of which can be gauged from Table 2.

Table 2. Pruning costs expressed as a percentage reduction for three stock types

Pruning height	Per-tree percentage reductiion in cost			
	Nonimproved seedling	Improved seedling (1st gen)	Juvenile steckling	Adolescent steckling
Low	100	– 16	– 20	– 30
Med	100	– 4	– 7	– 18
High (6.5 m)	100	– 4	– 6	– 16

It is also noted that the thinner branches of stecklings will frequently encourage pruning in excess of 6.0 m, further increasing the value of logs at rotation age. Additionally, the less butt sweep in the adolescent steckling plantations has been calculated to add a further NZ $ 500 per hectare to stumpage value (Arnold and Gleed 1985).

At TFL true comparative value studies have yet to be completed for adolescent steckling compared with seedlings. At 20 years of age one stand of adolescent stecklings from 8-year-old ortets have a clear quality advantage over seedlings of the same age, but stocking and silvicultural differences make an empirical comparison impossible. The frequently noted differences of bark thickness and reduced stem taper make it necessary for volume calculations to be made via stem section measurements rather than existing volume tables.

Stecklings taken from ortets into or beyond their adolescent growth phases enable improvements in stem and branch form that are associated with maturity to be expressed from the butt upwards, whereas in seedlings these traits are predominantly confined to the upper logs. Working against this strategy is the faster growth of today's genetically improved seedlings, especially on fertile sites, making it difficult and time-consuming to collect cuttings from the best (i.e., faster-growing) genotypes. In addition, the nursery "strike" rate rapidly reduces with increasing ortet age, e.g., cuttings from ortets 1–3 years old yield approximately 80–90% rooted stecklings, whereas those from 4–5 years can drop to 60%, and from 6–8 years donors, survival can fall to 40%.

Coupled with this situation is the improved juvenile and adolescent form of control-pollinated families that are becoming more readily available in New Zealand. Currently, TFL has temporarily forsaken the production of all stecklings from plantation ortets in favor of concentrating on other stock types (Table 1). It is anticipated that the next and perhaps most exciting opportunity for adolescent stecklings will be as the clonal stands reach the adolescent phase.

2.2 Juvenile Stecklings

Stool beds (hedge orchards) emanating from control-pollinated seed derived from the best progeny-tested parents available are used to vegetatively multiply the numbers of planting stock obtained from this often scarce seed source.

At TFL up to 1.0 million juvenile stecklings from the best genetic sources are used each year for forest establishment (Table 1). To date early growth differences in favor of stecklings have been observed within the forest stands between seedlings and juvenile stecklings. Where slower than average growth rates occur for stecklings, the cause is more frequently associated with the weaker root system of stecklings. For this reason strict quality controls are invoked at the time of lifting in the nursery, requiring all stecklings to have a minimum root structure (Fig 2). Subtle maturation differences exist between ramets taken from the same ortet in successive years.

Current TFL strategy is to maintain hedges for a 3-year productive period during which time each ortet is form-pruned to yield a maximum number of cuttings from a compact-shaped plant. At the completion of 3 years the hedges are replaced with further advanced genetic stock. The average total cutting yield per hedged plant for the 3-year period is 125.

Extensive trials with monthly production, as opposed to winter production only, have not been successful in economically improving the overall yields and as a consequence cuttings are taken and set only in autumn and winter.

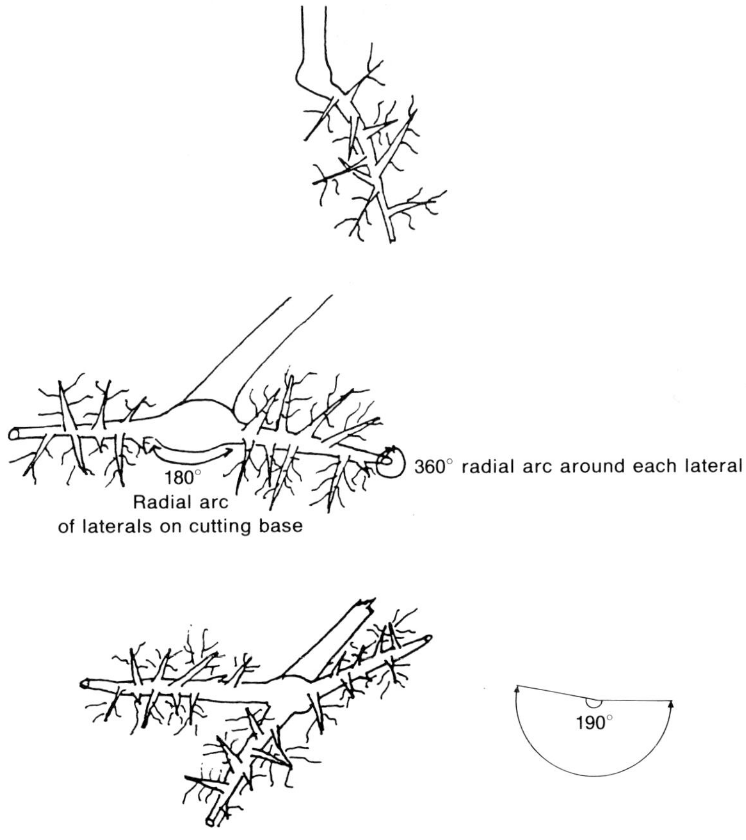

Fig. 2. Juvenile cuttings root systems. All three root configurations shown are acceptable

2.3 Plantlings via Tissue Culture

Clonal forestry is a major element within TFLs forest strategy, which in 1985 elected to use tissue-culture techniques for the clonal production of radiata pine planting stock.

Laboratory production is 3.0 million cloned plantlets per year (60 000/week). From this, 570 ha of clonal forest have been established and 1.0 million cloned plantlings are available for forest establishment in 1991. Using an average 600 stems per hectare (s/ha) for initial establishment, TFL will, in 1992, have the capacity to sustain 5000 hectares/year of clonal forest establishment.

At the commencement clonal testing was eschewed in favor of sustainable production from the best available seed sources with the object of as early as possible establishment of discrete clonal blocks (up to 50 ha). Representative cultures of all clones are periodically placed in the cool-store for an anticipated 2–3 years, by which time first clonal assessments within the blocks will be completed. At that stage poor performing clones will be dropped in favor of above average performers.

At TFL clonal blocks established to date have by necessity, due to varying ramet numbers/clone, been more varied in size than desirable. It was not until mid-1990 (5 years from commencement) that implementation of a clonal production strategy became possible due to increasing knowledge and experience with a large-scale tissue-culture system.

Adequate and balanced rooting of the tissue-culture shoots has been a major problem causing subsequent difficulties in the establishment of comparable clonal stocks. Thus of the 570 ha already established less than 50% will be used for clonal assessment.

Changes in greenhouse and propagation methods have now created better and more even planting stock. Concurrently, experience with a wide range of clones within both the laboratory and the greenhouse has been the basis of early rejection of many clones from the production program. Techniques have been developed that maximize in vitro rooting, leading to a dramatic increase in greenhouse survivals and subsequent fast and even nursery development.

3 Discussion

In order to maximize the return on forest investment, TFL seeks to be aware of and proficient in utilizing every opportunity presented during the establishment and maintenance of its radiata pine forests. Tree breeding, propagation, deployment, silviculture, and marketing are becoming more closely linked as systems both individually and collectively are continually refined and optimized. Breeding programs are being widened to better service family forestry through extensive control-pollination programs. Until breeding programs exist for clonal forestry, potential clonal "winners" are selected from within specific tested families.

3.1 Clonal Strategy

TFL has chosen 100 clones from the most genetically superior families available for the main production thrust. Concurrent with production and establishment of these clones is an intensive screening and testing program of 2000 clones, which will provide a source of more thoroughly tested and proven clones for deployment from 1995 onwards. The largest number of clones are established with minimal replication for initial field-performance screening. Semi-proven clones are then more rigorously evaluated with extensive replication in performance tests. From an early stage, potential superior clones enter the "production team" for deployment.

Optimum gains from clonal forestry come from the extensive deployment of good clones and learning how best to manage each for maximization of their specific traits and potential. It is considered that constantly deploying new clones which offer only limited additional margins of improvement will diminish real returns from clonal forestry. TFLs aim is to capitalize on efficiencies of "well-known" clones.

3.2 Forest Regimes and Clonal Deployment

Respective financial returns from clearwood and pulpwood logs are 8 and 2% respectively and hence there is no financial incentive to grow pulpwood crops per se. Thus TFLs objective is to maximize the production of clearwood logs.

The continuous improvement in both form and branch habit of genetically improved tree stock (stecklings and seedlings) has led to the gradual reduction in initial planting densities, from an average 1200 s ha in 1980 to 800 s ha in 1989. On many high index TFL sites and with high quality tree stocks this is reduced to 600 s ha.

These factors in turn will create a deficit in the volume of pulpwood logs that would normally be available from clearfell forest areas. It is presently calculated that such pulpwood arisings will diminish by as much as 70% and probably 90% for clonal forests (Gleed et al. 1991).

Final crop stockings for clearwood production are between 230–250 s/ha with approximately 300 stems removed in a pulpwood thinning between age 9–10 years.

In 1991 it is proposed to plant several clonal blocks with an inter-row mixture comprised of a fast-growing low-density clone for removal by row thinning for pulpwood with the remaining rows comprised of a higher than average density clone for the final crop.

No estimate is made of the increased economic value of this strategy but it is expected that the increased uniformity and improved mechanical groundwood characteristics of the pulpwood clone will provide some financial premium as stumpage.

By the year 2005 TFLs forests could yield in excess of 400 000 tonnes of purpose-grown pulpwood emanating from clones chosen for desirable pulp-making characteristics (Fig. 3).

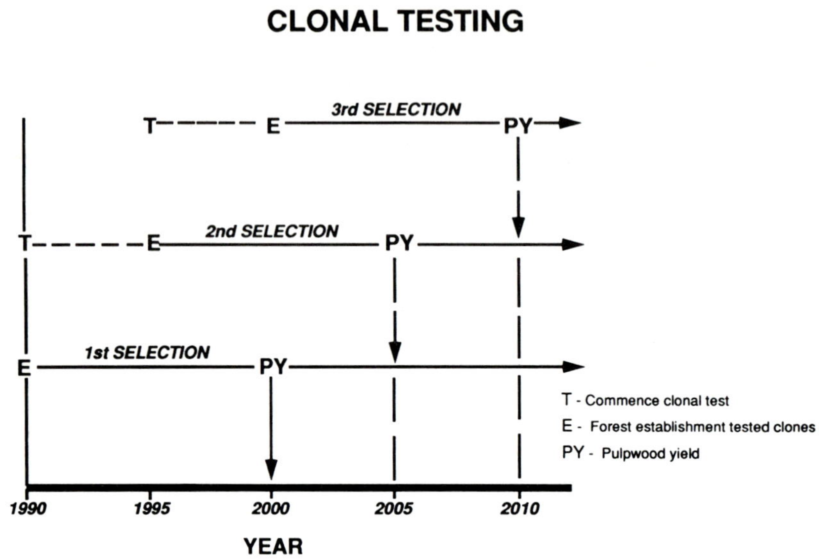

Fig. 3. Schematic TFL clonal testing and establishment program

3.3 Financial

Vegetative propagules are more expensive (up to five times) than seedlings, and tissue-culture plantlets have been more expensive than adolescent stecklings. In 1991, TFL tissue-culture plantlets will be produced at less than the budgeted costs of adolescent stecklings and, as recent new developments in tissue culture take operational effect, the cost of plantlets will fall further (note all are direct costs only). Nevertheless tissue-culture plantlets are likely to remain dearer than both seedlings or juvenile cuttings (Fig. 4).

Real costs for seedlings will marginally increase over time with the increasing cost of genetically improved seed especially from control pollinations. Adolescent stecklings will fluctuate in real costs dependent upon the location and distance of suitable donor plantations from the nursery. No significant improvements in cost efficiencies of collecting, setting and growing are foreseen, except to improve the "strike" rate above the maximum of 60% currently achieved from 6–8-year-old ortets. Tissue-culture plantlet costs are destined to fall for some time yet as the TFL unit comes up to full production with all clones having a high capacity for in vitro rooting.

Whilst cost efficiencies are always paramount in every aspect of TFL management, the initial high cost of cloned tissue-cultured plantlets has not deterred TFL from progressing to full operational scale. Several factors help influence this decision; they include:

1. The cost of planting stock is not very influential in the final analysis of radiata forest investment.
2. The expected returns from clonal forestry when fully optimized fully justify the initial investment and increased costs of planting stock.

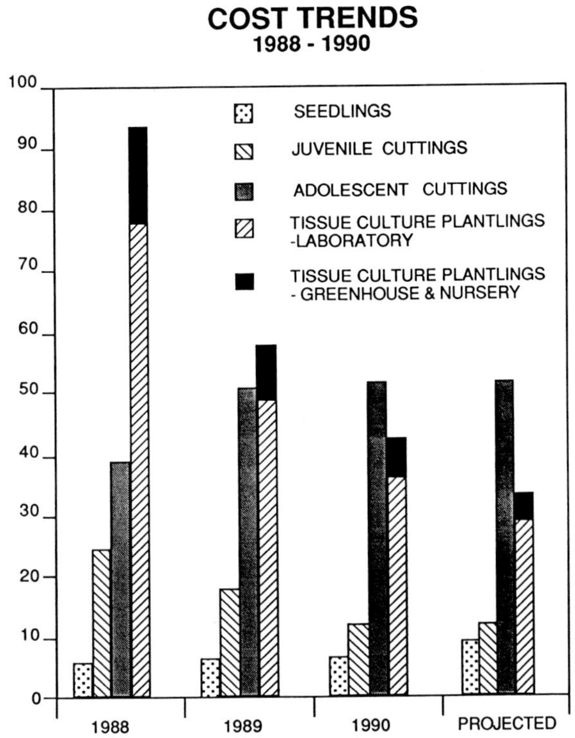

Fig. 4. Production cost trends for four stock types at TFL

3. TFL forest accounting policy takes as "profit" the annual increase in volume of the growing forests. Thus the influence of tree improvement is acknowledged at an early stage of the forests rotation.
4. The willingness to be innovative in order to create opportunities for improved profit performance.
5. Recognition that skills experience and scale will influence unit costs downward over time.

Finally, TFL has the belief that the most successful forest companies in 2000 and beyond will be those that understand and practise elements of forest biotechnology as an integral part of forest management.

References

Arnold RJ (1990) Control-pollinated radiata pine seed—a comparison of seedling and cutting options for large scale deployment. NZ For 35.3:12–17

Arnold RJ, Gleed JA (1985) Raising and managing radiata pine cuttings for production forests. Aust For 48.3:199–206

Gleed JA, Arnold RJ, Siegfried J (1991) Radiata pulpwood production – an optimal approach via clonal forest plantations. APPITA 44:323–324

9 Arguments for Redwood Clonal Forestry

J.A. Rydelius[1] and W.J. Libby[2]

1 Introduction

Some concerned citizens are cautious about clonal forestry because they worry about "fooling Mother Nature". But Mother Nature normally clones coast redwood (*Sequoia sempervirens*). Stump sprouts, or sometimes stem sprouts or root sprouts or adventitious shoots, produce multiple copies of successful clones in local clumps for two, three, or even more generations, occupying that site for centuries or even millenia (Fig. 1).

2 Natural Cloning by Sprouts

Although clonal forestry is a relatively new concept as far as its operational application is concerned, coast redwood clones have been an important, albeit inadvertant, method of regeneration following redwood harvests or after natural disasters such as windthrow and fire.

Soon after germination, all redwood seedlings form a cluster of two buds at the junction of their root and shoot. These buds, which otherwise remain dormant, divide to form a basal burl as the seedling grows (personal observations, JAR, 1960–1990). In several hundred years, this basal burl may consist of thousands of live buds and weigh 3–5 metric tons or more. The buds of this burl system have the capability of sprouting at any time that they are stimulated to do so, although the propensity for sprouting tends to decrease among older and/or larger trees. Factors that stimulate the sprouting of burl-system buds include fires that kill and even consume the aerial portions of seedlings or trees, windthrow, sudden exposure to sunlight, and, during the most recent period of redwood's evolution, tree felling by *Homo sapiens* for harvest or other purposes.

Almost without exception, seedlings and young saplings sprout following damage or other stimulation. They produce relatively few sprouts, because only a few buds have proliferated from the original two. When larger trees are in a thrifty condition, clumps containing several hundred sprouts are common

[1]Simpson Timber Co., PO Box 1169, Arcata, CA 95521, USA
[2]Department of Forestry and Resource Management, Mulford Hall, University of California, Berkeley, CA 94720, USA

Clonal Forestry II, Conservation and Application
Ed. by M.R. Ahuja and W.J. Libby

Fig. 1. A typical circle of redwoods growing around an old stump. Note the similar bark characteristics and branch angles. While seedlings sometimes also germinate and grow successfully in the favorable microsite of a stump-edge, the consistencies within such clumps support the general notion that most members of such groups are clonal descendants of some past seedling ortet.

following timber harvest or other sudden killing of the main stem. Very large and very old redwoods, which may be in a less vigorous condition, often produce only a few sprouts, and sometimes none upon logging or death of the tree. Reasons for failure to sprout may be excessively thick and thus impenetrable bark, earlier damage to the basal burl, and/or because the live buds have been overgrown by woody tissue.

In addition to the proliferation of the buds in the basal burl, redwood trees produce axillary buds along the entire length of the bole and branches. These buds typically elongate only enough each year to maintain their terminal meristems just below the cambium layer. They are stimulated to sprout by the same factors that stimulate the sprouting of basal-burl buds.

Sprouts have been seen to arise adventitiously from wound tissue on cut or broken stems or branches. Sprouts from injured or exposed roots are also occasionally observed. Because of the development of such adventitious buds and, more important, because of dormant axillary and basal-burl buds, vegetative reproduction of redwoods has occurred since before *H. sapiens* intervened in natural redwood ecosystems. It is clear that multiple copies of successful clones in local clumps have resprouted for several clonal generations, thus occupying local areas of increasing size for very long periods of time.

3 Natural Reproduction from Seeds

The natural clonal reproduction of redwood is so prevalent that many casual observers in the past (and perhaps today) were of the opinion that redwoods could not reproduce from seeds. Such is clearly not the case. Redwoods do produce and reproduce from seeds, although much more frequent and abundant seed crops are produced by trees in the northern, eastern, and southern portions of the native range than in its center. The viability of the seeds also varies, with some regions having no open-pollinated families germinating in excess of 5%, while o-p families in other regions routinely have about 40% germination, and some have in excess of 60%.

Once young-growth trees are sexually active, they generally produce larger cones than do nearby old-growth trees, and the seeds from the young-growth trees usually germinate in higher percentages. Recent experience with helicopter collection, picking cones near the tops of young-growth trees, indicates that seeds from cones near the tops have higher germination than do seeds from cones lower in the crowns. Both of these observations may have importance with respect to the generational makeup and neighborhood size of naturally regenerated redwood forests.

It is a paradox that redwoods, once established, are among the least likely of tree species to be killed by insects and disease, but redwood germinants and first-year seedlings are unusually susceptible to soil pathogens (Muelder and Hansen 1961; Hansen and Muelder 1963; Davidson 1971). With rare exceptions, redwood seeds must germinate on exposed mineral soil to become successfully established as seedlings. Their successful establishment is enhanced if a moderate fire has killed some or most of the pathogenic fungi near the soil surface, and it is further enhanced if the seeds germinate in forest openings.

Unfortunately for redwoods, the conditions that favor redwood seedling establishment are also favorable for the establishment of associated species. These associated species are subordinate to redwoods in mature redwood forests, but as juveniles some of them grow rapidly, suppressing and dominating redwood seedlings for periods of 40–50 years or more. Under dense hardwood stands, particularly of alder, redwood seedlings will linger until the alder stands break up. If conifers such as Douglas-fir, Sitka spruce, and western hemlock are present, they will also respond to the breakup of the hardwood overstory and do so with faster growth than that of the redwoods. In many cases, the redwoods do not recover from this suppression, and never grow through to become dominant if the other species are sufficiently overtopping.

On the other hand, redwood sprouts begin their growth within 4–6 weeks after the original seedling or tree is damaged. Because of their established root systems, if other species are also eliminated by whatever caused the damage, the redwood sprouts will usually outgrow newly established seedlings of competing species. This effect is particularly evident if the harvest or other cause of damage occurs part way into the growing season. Most seedlings successfully germinate and establish only during a short period in the spring, while redwood sprouts can begin vigorous growth at times unfavorable to seedlings.

4 It Is Hard to Kill a Healthy Redwood

Because of the natural sprouting of redwood, clonal regeneration of redwood has been an inadvertant result following the commercial harvest of redwood. It has not, however, always been an objective of management. In the early days of redwood harvesting, 80–100 years ago, it was often the objective of owners of redwood forests to remove new forest growth and establish grass for grazing purposes. Such conversions of land use were typically attempted by burning the logging debris, followed by sowing grass seeds over the temporarily cleared land. Although such efforts usually resulted in the establishment of grass, tree seedlings and particularly redwood sprouts soon occupied substantial areas in the cleared pastures, requiring repeated chopping and reburning of the young trees. The redwood stumps kept sprouting, and were joined in this by sprouts from the occasional or sometimes abundant redwood seedlings recruited to the site. In this manner, both the absolute and relative frequencies of redwoods increased following each attempt to maintain the grass. A common result was that the discouraged grazer gave up, and the forests that then grew on such sites had a higher percentage and density of redwoods than existed in the original stand, or that would have existed if site conversion had not been attempted.

5 The Composition of Redwood Forests

Because of the size of mature redwood trees, natural redwood forests normally consist of relatively few redwood trees (and even fewer clones) per unit area, usually mixed with large trees of other species. Twenty-five large redwood stems plus 25–30 large stems of other conifers per hectare is about average, with substantial variation around those numbers. In managed second- and third-growth redwood forests, the number of stems per unit area is increased to about 450 total (redwood and other) stems per hectare at harvest, which occurs at about 40–80 years from establishment.

In the past, an increase in redwood stocking following logging occurred, at least in part, for reasons described above. For the last 15 years, an additional increase in redwood stocking has occurred as a result of planting nursery-grown redwood seedlings to supplement naturally occurring redwood sprouts and seedlings. Because redwood forests are almost always a mixture of redwoods with other conifers (primarily Douglas-fir) and a few hardwoods, regeneration foresters usually plant a mixture of conifer species in an effort to supplement the natural regeneration, and to maintain a mixture of tree species in the managed stands similar to that in the native stands.

With respect to the redwood component of the regenerated forest, interplanting seedlings increases the number of unique genotypes per local area, compared to the mosaics of small clones that characterize native redwood forests. Preliminary research (Millar et al. 1985) as well as research in progress (pers. commun. Thimmappa Anekonda, Dept. Forestry, UC Berkeley, 1990) indicates that the genetic variation among redwood stands within a physiographic area is small compared to genetic variation within stands. Since the seeds for redwood reforestation are carefully collected from the local physiographic region, greatly increased genetic variability is not likely as a result of this interplanting. However, the gaps between genotypes (biotypes) must be much smaller among 200–300 redwood stems per hectare, many of them single-copy seedlings, as compared with the situation in native stands where fewer than 25 redwood genotypes per hectare or neighborhood is the norm. This perhaps poses some increased risk of cross-adaptation and epidemic spread of short-lived pests and pathogens (Libby 1982). Compared with other forest tree species, redwood is known to be on the low end of the range of susceptibility to epidemics of pests and pathogens. Whether this is due, at least in part, to the relatively few biotypes per unit area and their clonal-mosaic distribution in most native stands of redwood is an interesting question.

6 Redwood Clonal Forestry

The operational production of clonal redwood planting stock provides an opportunity to maintain the wider genetic gaps that exist in natural redwood forests while achieving the desired proportion of redwood stems in managed redwood forests. The use of clonal redwood planting stock provides the additional advantages of increasing forest growth and yield, as clones can be selected for such traits as growth rate, bole form, branch size and angle, and wood properties, in addition to their selection for effective vegetative propagation and early field survival.

6.1 Plantlings

Redwood was the first conifer to be successfully tissue-cultured (Ball 1950), although independent plantlets were not produced until later. During the last two decades, the AFOCEL group at Nangis and Paris have pioneered the use of tissue culture for rejuvenating mature tissue, using redwood as a particularly useful species for their studies (Franclet et al. 1987). The Simpson Timber Company Tissue Culture Laboratory at Korbel, California, is among the first to put the process into operational practice. In Korbel, several redwoods in excess of 300 years old have been sufficiently rejuvenated to be entered in a current 200-clone provenance test (Kuser et al., unpubl. 1988); near-complete rejuvenation has been accomplished with tissue from sprouts and branch meristems of redwood trees in the age range 40–80 years. In at least one case, the tissue-culture-rejuvenated plantlings from trees in the 40–80-year range appear to be more juvenile than do seedlings of the same population planted with them (pers. observ. 1978 redwood seedling-plantling-steckling comparison at Russell Reservation, by WJL during 1978–1990).

This latter technique, using early-mature (40–80 years old) trees selected for growth and form in natural stands, is the basis for the current clonal-forestry program at Simpson Timber Co. As of 1990, 244 plus trees have been selected. Tissue collected from 130 of these plus trees is being or has been put through a rejuvenating tissue-culture step. Rejuvenated plantlets are produced in sublines of the tissue-cultured clones. After successful transfer to a greenhouse, the resulting plantlings are hedged and most are planted in a production hedge-orchard (Fig. 2). These hedged plantlings produce operational quantities of cuttings, which are rooted for both testing and reforestation purposes.

Sixteen tested clones are now in routine operational production at Simpson's Korbel hedge-orchard, and these 16 clones supplied 65 000 stecklings of the 1 000 000 propagules planted by Simpson in 1990. Projections for 1991 indicate that 170 clones will be in tissue culture and 20 will be in production, producing 125 000 stecklings. The number of clones in tissue culture and undergoing various phases of juvenility and candidacy testing will peak at 250 in 1993, and then decline to about 150 as clones are culled from the program. By 1995, annual steckling production will provide the full million or so propagules needed for all redwood planting by Simpson, from about 130 clones that have performed well in early testing.

A similar program, using tissue from selected redwood trees that have survived and grown well for many decades in French parks and gardens, has

Fig. 2. These plantlings are growing in a portion of the Simpson hedge-orchard. The clones are planted contiguously in rows, and clonal differences may be seen between the two foreground rows

provided cuttings for commercial nurseries and industrial use in Europe since 1978 (Franclet et al. 1987).

Additionally, adolescent redwood trees of outstanding performance are being selected in native stands or in young plantations originating from wild seeds (i.e., not from the seed orchard). These are also being cloned and entered in the Simpson clonal-testing program and the breeding program. Increasingly, new clonal selections will be from fully pedigreed stock selected from advanced-generation tests in the breeding program.

6.2 Breeding

Redwood, perhaps because of being hexaploid, does not exhibit severe inbreeding depression under reasonably favorable conditions. But growth and survival of inbred redwoods is substantially less than that of related outcrosses in more stressed conditions (Libby et al. 1981; Kirchgessner and Libby 1985). Thus, in addition to normal testing protocols (see Frampton and Foster, Chap. 9, Vol. 1), unpedigreed young clones from open-pollinated families in nature must be tested for inbreeding depression in moderately stressed field conditions. The inbreeding problem is eliminated by selecting clones from pedigreed outcrossed families.

For tree-improvement purposes, the natural range of redwood has been divided into a number of breeding zones. Within each breeding zone, redwood plus trees have been selected for entry in breeding-and-seed-orchards. At present, it is tissue from these plus trees that is being rejuvenated in tissue culture for propagation for testing and production plantations.

Some cuttings are rooted from each of the selected near-harvest-age redwoods without going through the tissue-culture step. These root in low frequency, but develop as mature stecklings that soon produce both male and female strobili. They are deployed to breeding orchards where, in 1990, 227 (full-sib and reciprocal) crosses were made among 35 clones as females and 87 as males (some as both) to produce pedigreed juvenile progeny. Since redwood is unusual, perhaps unique, in that both mitochondrial and chloroplast DNA is inherited through the male line (Harrison and Doyle 1990), particular attention is being paid to the breeding design and to the male contributors to these pedigreed families. Approximately 120 genotypes from each full-sib family have been or will be hedged and tested for future deployment in the redwood clonal-forestry program.

6.3 Stecklings

As indicated above, redwood seeds germinate at variable and generally low percentages. Thus, in order to achieve efficient spacing in nursery beds or efficient utilization of containers in greenhouses, redwood nurseries must pre-germinate redwood seeds and plant by hand, or oversow and thin, both costly operations. Cuttings can be taken from hedges and set in the nursery at similar costs. Germinating redwood seedlings are highly susceptible to several damping-off fungi, and gray mold (*Botrytis*) is sometimes devastating on seedlings in the nursery. By contrast, recently rooted cuttings of redwood rarely suffer from these diseases, another advantage of stecklings over seedlings.

Cuttings from the pedigreed juvenile hedges and from rejuvenated hedged plantlings typically root in excess of 95%, and do so within 10 weeks of being set. A juvenile hedged steckling or seedling provides 10–100 cuttings in its first year, depending on growing conditions and intensity of cutting harvest, and it can provide hundreds of cuttings in each subsequent year. Hedged rejuvenated plantlings follow this same schedule after a delay of about 1 year for testing of the plantlets for juvenility. Thus, hedge-origin cuttings are providing a cost-

Fig. 3. The six redwood trees in the foreground are three clonal pairs. Each of three seedlings was cloned by rooting a cutting from it, and the ortet and ramet were then planted side by side. As of this 1990 photo, these trees had been on the site for 22 years. In each pair, the seedling is on the *left* and the steckling on the *right*. The basal swelling typical of seedlings is visible on the *left-most tree*. Otherwise, the differences between members of each clone are small (Jiang 1982), and the differences among clones are large. Testing is superfluous to decide not to deploy the clone on the right, which is not only relatively short, but has multiple ramicorn branches on both the ortet and ramet. When this plantation was clonally thinned in 1978, the stumps of the seedlings had more and lower sprouts than did the stumps of the stecklings, but the sprouts from both kinds of stumps were clearly sufficient to keep each redwood, whether seedling or steckling, occupying its site in the plantation. (Photo by J. Morse)

effective step between the rejuvenation of selected redwood genotypes, or the production of pedigreed redwood genotypes, and their deployment to tests and to production plantations. At present, Simpson's strategy is to favor tested clones over the vegetative multiplication of pedigreed genotypes tested at only the family level.

There is convincing evidence that, as in so many other conifers (see Greenwood and Hutchison, Chap. 3, Vol. 1), there are large differences in growth and form between redwood stecklings from juvenile and mature ortets, and also among stecklings from different parts of mature redwood trees (Tufuor 1973). However, a study of six plantations containing redwood seedlings and stecklings from juvenile ortets, with one exception, failed to find consistent or statistically significant differences between observed and measured traits of these two kinds of propagule (Jiang 1982). The one exception was basal swelling; if a substantial difference existed between a seedling and its juvenile-steckling ramet, the steckling consistently had less basal swelling than did the seedling, an effect probably associated with the basal burl (see above).

Less information is available on adolescent stecklings as compared to either juvenile stecklings or seedlings. There is some evidence that adolescent stecklings have fewer basal sprouts and multiple stems than do seedlings or rejuvenated plantlings (Jiang 1982; pers. observ. 1978 redwood seedling-plantling-steckling comparison at Russell Reservation, by WJL during 1979–1990). The frequency of multiple stems is a trait of silvicultural importance that could favor deploying stecklings at a maturation state greater than fully juvenile (see Gleed, Chap. 8, Vol. 2).

At present it seems safe to conclude that juvenile stecklings perform similarly to seedlings (Fig. 3). It is still too early to tell whether rejuvenated plantlings and their hedge-origin stecklings will be similar to seedlings, or whether they may even have some advantages over seedlings. It is also still too early to tell whether clones can be maintained at a particular maturation state by appropriate hedging, or by repropagation from tissue culture. Meanwhile, sublines of the selected and rejuvenated tissue-cultured clones are maintained in tissue culture, and new hedges for producing cuttings for production plantations are begun from fresh plantlets induced in these tissue-culture sublines every 3–5 years.

6.4 Operational Deployment

Reforestation with redwood clonal planting stock is operationally similar to reforestation with seedlings with respect to grading, packing, storage, shipping, and field planting. The decision as to how many stecklings of each clone to produce is based on the harvest and site-preparation plan developed for each year. Prior to the time that cuttings from the hedges are stuck, it is generally known where and how much land will be ready for planting during the next planting season. Based on this information, and on other information such as the density at which the areas are to be planted, the species mixture to be planted, and the stocking of redwood anticipated as the result of sprouting redwood stumps (natural regeneration by wildling redwood seeds is too unreliable to be used in this calculation), the desired number of stecklings of each clone is determined. Which single clone or mixture of clones is to be planted in each area is determined by comparing the planting-site conditions with the site conditions from which the clones were originally chosen. Increasingly, this origin information will be supplemented with data accumulated from the clonal tests.

Redwood clonal planting stock can be produced in any of a number of stock types. Most often, it is produced as containerized stock with a tapered root plug about 4 cm diameter at its top and from 15 to 20 cm long. Such "plugs" are a 1-year crop in the greenhouse, with cuttings taken from hedges during November-February and with stecklings ready for outplanting anytime during the following November through February. Less commonly used are "plug-ones", which ordinarily require 1 year of greenhouse growth in a container followed by a 2nd year of growth bareroot in an outside nursery bed. Plug-ones, planted as bareroot stock, are used on difficult sites or where rapid commencement of unchecked growth from relatively large stock is desired.

6.5 Expected Gains

A review of data for the redwood plus trees that have been selected to date indicates that their volumes exceed those of nearby trees of the same age, including the check trees, by 50–400%. Clearly, with an averaging out of environment, with the likely higher inter-tree competition in plantations made up of such plus clones, and with substantial proportions of most planting sites sometimes being occupied by wildling seedlings and often by sprouts of redwoods, by wildling and planted seedlings of Douglas-fir, and by seedlings and sprouts of other species, operational volume gains of that magnitude cannot be expected. As an upper-bound comparison, selection and clonal deployment of eucalypts in Brazil has resulted in volume increases in the neighborhood of 60–110% (see Zobel, Chap. 7, Vol. 2); the clonal redwood plantings should be able to capture some reasonable proportion of such figures. With improved bole form and branching characteristics, increases in harvest index of the selected redwood clones may be as important as increases in volume per unit area.

7 Conclusions

We conclude by returning to our opening point. Most redwoods sprout when harvested or when they are otherwise injured or their tops killed. And, as ranchers and right-of-way clearers have found to their dismay, redwoods once established are very hard to kill. This means that once a redwood tree occupies a site, whether planted or by natural means, it is capable of occupying that site for a very long period of time; often as a naturally developed clone. As long as we are planting redwoods on productive sites, it makes sense to plant clonally tested good ones.

References

Ball E (1950) Differentiation in a callus culture of *Sequoia sempervirens*. Growth 14:295–305

Davidson JGN (1971) Pathological problems of redwood regeneration from seed. PHD Diss, Univ Cal, Berkeley, 288 pp

Franclet A, Boulay M, Bekkaoui F, Fouret Y, Verschoore-Martouzet B, Walker N (1987) Rejuvenation. In: Bonga JM, Durzan DJ, (eds.) Cell and tissue culture in forestry, vol 1. Nijhoff, The Hague Dordrecht, pp 232–248 (422 pp)

Hansen JH, Muelder DW (1963) Testing of redwood seed for silvicultural research by X-ray photography. For Sci 4:470–476

Harrison RG, Doyle JJ (1990) Redwoods break the rules. Nature (London) 344:295–296

Jiang IB-J (1982) Growth and form of seedlings and juvenile rooted cuttings of *Sequoia sempervirens* and *Sequoiadendron giganteum*. MS Thesis, Univ Cal, Berkeley 200 pp

Kirchgessner KA, Libby WJ (1985) Inbreeding depression in selfs of redwood: Rooting. Cal For For Prod 60:2 pp

Libby WJ (1982) What is a safe number of clones per plantation? pp 342–360. In: Heybroek HM, Stephen BR, von Weissenberg K (eds) Resistance to diseases and pests in forest trees. PUDOC Wageningen, pp 342–360 (503 pp)

Libby WJ, McCutchan BG, Millar CI (1981) Inbreeding depression in selfs of redwood. Silv Genet 30:15–25

Millar CI, Dunlap JM, Walker NK (1985) Analysis of growth and specific gravity in a 20-year-old provenance test of *Sequoia sempervirens*. Cal For For Prod 59:8 pp

Muelder DW, Hansen JH (1961) Biotic factors in natural regeneration of *Sequoia sempervirens*. In:Proc 13th IUFRO Congr, Vienna, pt 2, vol 1 (24–4/1):15 pp

Tufuor K (1973) Comparative growth performance of seedlings and vegetative propagules of *Pinus radiata* (D. Don) and *Sequoia sempervirens* ((D. Don) Endl.). PhD Diss, Univ, Berkeley, 207 pp

10 Clonal Forestry with Tropical Hardwoods

A.F. Mascarenhas and E.M. Muralidharan[1]

1 Introduction

The tropical forests are resources of enormous potential value as fuel wood, timber, and pulpwood. Although half of the forests of the world are in the tropics, the rate of depletion of these forests is a cause for concern.

Interest in plantation forestry in the tropics has picked up only in the last few decades. Tree improvement programs on the lines of what has been developed for temperate hardwoods and conifers have been initiated in many countries where forest land is a valuable asset. In these countries it is of vital importance to maximize forest yields to solve pressing problems with regard to the shortage of fuelwood, timber, and pulpwood.

Natural tropical forests exhibit considerable variation in the composition of species, and in the size, age, and form of trees within the same area of the forests (Brazier et al. 1976). This heterogeneity presents intractable problems for commercial exploitation and application of silvicultural treatments. The removal of the preferred species, which form only a small fraction of the total, thus results in the destruction of the whole forest. With the increased demand for industrial wood and pulpwood, requiring particular wood characteristics, large-scale artificial plantations of selected tree species will be required.

Tropical tree species typically have a low population density, a variety of breeding systems and a relatively greater inter-population variation than temperate trees. The lack of basic information about the genetics and reproductive biology of most tropical trees hampers tree-breeding programs in the tropics (Bawa 1976). Most of these programs have progressed only as far as the selection of plus trees and establishment of seed orchards (clonal or seed-raised). In many tropical tree species low seed setting, short seed viability periods, higher incidence of seed predation, and inaccessibility to wild populations further add to the problems encountered in tree breeding. In this situation, vegetative propagation and clonal forestry assume importance for bringing about quicker, short-term tree improvement by capturing the existing genetic variation directly.

Zobel and Talbert (1984) have summarized the application of vegetative propagation in forestry for germplasm conservation, genetic research, breeding, and operational planting.

The need for preserving an appropriate sample of germplasm in clone banks is urgent in cases of many tropical tree species that face total destruction of their

[1]Division of Biochemical Sciences, National Chemical Laboratory, Pune 411 008, India

Clonal Forestry II, Conservation and Application
Ed. by M.R. Ahuja and W.J. Libby

wild habitat. While in situ conservation in the natural state is best for some purposes, several tropical tree species are endangered in the wild due to regeneration problems like poor seed set, poor viability of seed and seedlings (e.g., *Vateria macrocarpa*) or sterility (e.g., *Vernonia shevaroyensis*).

In provenance testing, a more efficient, clearer evaluation of the genotype x, environment interaction is possible, using clones simultaneously under different agroclimatic conditions.

Progeny testing of plus trees requires an adequate quantity of seeds produced by open or controlled pollination. This can be achieved by assembling clones of plus trees in clonal breeding orchards, where intensive management improves the flowering and seed formation. The smaller size of the trees at the flowering age makes controlled breeding easier.

The interest in the use of rooted cuttings as propagules in operational (production) forestry of tropical trees is growing with an increasing number of tropical species becoming amenable to vegetative propagation (Leakey 1987). Cloning produces plantations of greater uniformity than is possible from seed. More than double the genetic gain for growth and volume appears to be possible, e.g., *Eucalyptus* (Campinhos and Ikemori 1980; Zobel et al. 1982).

In this chapter on tropical hardwoods, an account is given of the various approaches to clonal propagation of three forest tree species viz. teak, tamarind, and bamboo. Tree improvement and large-scale propagation in these important tree species have been hampered by several limitations in conventional breeding and propagation methods. In the account that follows, these problems and the methods adopted to overcome them will be discussed. The various methods of vegetative propagation developed for these species and their relative merits and demerits are compared. A general description of the distribution, biology, and economic importance is also given under each of the tree species. This discussion depends mainly on the information available on the work carried out in India. Micropropagation of tropical trees has been one of the major subjects of tissue culture research at the National Chemical Laboratory (NCL) at Pune in India for the past decade, and the information presented here is based on the results published over this period.

2 Teak (*Tactona grandis;* Verbenaceae)

Teak is one of the most important timbers of the world. Teak combines qualities like durability, workability, and resistance to termites, fungi, and weather. It can be used for a wide range of purposes, which includes furniture making, construction, and ship building.

2.1 Distribution

The natural geographic range of teak is southeast Asia (India, Burma, Thailand, and Indonesia). Two other species of *Tectona* viz. *T. hamiltoniana* and

T. philippinensis are endemic to parts of Burma and Philippines respectively. As an exotic, teak has also succeeded in other areas of the tropics (Evans 1982).

2.2 Climate

A moist, warm tropical climate is most favorable for teak, although it grows under a wide range of climatic and edaphic conditions, and can also grow in areas with an annual rainfall as low as 60 cm and as high as 500 cm (Hedegart 1976; Anon 1976b). Teak prefers rich soils with good drainage and depth, as in the lower slopes of hills, and also thrives on freshly deposited alluvia. Teak forests have been classified into the following five types based on the annual rainfall, by Seth and Khan (1958): very dry (less than 90 cm), dry (90–130 cm), semi-moist (130–165 cm), moist (165–250 cm) and very moist (over 250 cm).

2.3 Biology

On favorable sites teak attains a large size, with a tall clean and cylindrical bole. Older trees can develop a fluted and buttressed base.

Teak leaves are large, obovate, and have an opposite decussate arrangement. Flowers occur in terminal panicles and are small, whitish, actinomorphic, and hermaphrodite. Teak is insect pollinated and is mainly a cross-pollinating species, self-incompatibility being 96–100% (Hedegart 1976).

Teak fruits develop to full size 50 days after pollination but ripen only after 120–200 days. The fruit is a drupe with a tetralocular, hard, and bony endocarp.

2.4 Timber

Teak has a worldwide reputation as a quality timber, both for its mechanical properties and its appearance. It is one of the most durable of woods. The presence of substituted anthraquinones and desoxylapachol in the wood is responsible for its termite and fungal resistance (Anon. 1976c). Teak is easily air-seasoned and is easy to saw and work.

The best teak timber comes from Burma and Thailand. In India, teak from the south is considered equivalent in strength to that from Burma.

2.5 Provenances

Striking variations are observed between populations of teak in different locations, especially in India (Hedegart 1976; Kedharnath and Mathews 1962). The timber from dry zones exhibits good figure and is valued for paneling and furniture. Teak from Allapalli in the center of India has a reputation for high wood quality. The "Teli" teak grown in Karnataka in the south of India is known for stronger and more durable wood, a shorter leafless period, better seed germination, and resistance to the leaf skeletonizer.

There is evidence that much of this variation is inherited (Kedharnath and Mathews 1962). Kumaravelu (1979) studied the isozyme patterns of four clones of teak and concluded that isozyme analysis can be used to identify individual clones.

2.6 Pests and Diseases

Teak trees are attacked by over 50 insect pests, the most serious in India being the teak defoliator (*Hyblacea purea*) and teak skeletonizer (*Hapalia macheralis*), the larvae of which often destroy the entire foliage of a tree. This has a detrimental effect on timber formation and seed production. Inflorescences are damaged by the tree hopper (*Macherota* spp; Homoptera). The teak beehole borer (*Xylentes ceramicus*) also causes severe damage in plantations in Thailand and Burma. In India, damage by the angiosperm parasite *Dendrophthoe* is serious in plantations. Rot and decay of heartwood is caused by *Peniophora rhizomorsulphurea*, *Polyporus zonalis* and *Fomes lividus*.

2.7 Teak Improvement

Genetic improvement programs have been initiated in most of the teak-growing countries in the past few decades. Provenance trials have established the patterns of genetic variability available in various teak zones for the useful traits. In India, a breeding program formulated by Kedharnath and Mathews (1962) has been widely adopted by the Forest Departments in the different states of the country. Large numbers of plus trees have been selected in these states and seed-orchards established by various methods of vegetative propagation like grafting and budding.

Selections of plus trees are based on the following main traits in India (Kedharnath and Mathews 1962).

1. Superior growth in height and diameter.
2. Stem free from spiral grain and excessive fluting.
3. Resistance to frost in the north and to drought in the south.
4. Resistance to the leaf defoliator and the leaf skeletonizer.
5. Ability to produce wood with desirable characteristics.
6. Ability to produce regular crops of viable seeds.

These identified plus trees are then included in the National Register.

Seed orchards are established in each teak-producing zone and at sites where teak grows well, bears flowers regularly and is separated from large stands by at least 400 m. The minimum extent of a seed orchard is 10 acres and a minimum of 30 parent trees is included, represented by several grafts from each. Layouts with a special type of randomization that permits equal chance for each clone to be pollinated by all the other clones are designed. Initial spacing in teak seed orchards is 4×4 m, increased to 8×8 m after about 10 years, and finally to 16×16 m (Kedharnath and Mathews 1962).

2.8 Regeneration

Teak regenerates from seed but several factors influence the extent of natural regeneration. Nanda (1962) established that flowering in teak is dependent on the amount of available light, as lower branches inaccessible to sunlight do not flower. Development of fruits is also similarly affected. Attack by the tree hopper and the defoliator results in the damage of the inflorescence, which also affects fruit production.

Seeds show poor germination in several teak-growing provenances. Temperature plays an important part in inducing germination when sufficient moisture is available. However, exposure to sun or to alternate soaking in water and scorching in the sun improves germination. Natural regeneration of teak is also influenced by soil pH (6–8 being favorable), and by the incidence of forest fires, which is favorable in moist zones, but harmful in the dry zones. Availability of light, soil aeration, weed growth, and grazing are the other factors influencing natural regeneration.

2.9 Artificial Regeneration

Plantations of teak were established in India as early as in 1844. Various methods to hasten and improve germination of teak seeds are used in different teak areas (Anon. 1976c).

A common method is to alternately soak the fruits in water and dry them in the sun till the hard testa is softened. This weathering may take few days or over a month. In the Bengal pit method, seeds are placed between layers of leaves in a sealed pit watered through perforated bamboo tubes. This method requires about 10 days to be effective. Scorching the seeds in a light fire or hot sand or subjecting it to bacterial action and acid or alkalis are the other techniques.

Mechanical extraction of seeds from the hard testa has also been found to improve germination. Dabral (1976) described a method of extracting seeds using a modified nut cracker with a 56% germination efficiency.

Teak plantations are raised by transplanting nursery grown seedlings directly or after preparation of "stumps", which were first used during the 1920s. Stumps are prepared from 1–2-year-old nursery-grown seedlings having a thickness 1–2 cm. The seedlings are trimmed retaining 1.5 to 5.0 cm of the shoot and 15–25 cm of tap root. Stumps split into two halves can also be used. It is now a common practice in all teak-growing areas, as it offers the advantage of easy transportation from nursery to site with little damage.

Stumps are planted after the first showers in crowbar-prepared holes in areas with good rainfall or in pits, refilled with soil, in other areas. Pre-sprouted stumps from beds or polythene bags are used where heavy weed growth is expected. Regular weeding is required in the early stages of seedling growth.

2.10 Vegetative Propagation

The assemblage of clones of plus trees in seed-orchards constitutes an important step in teak-improvement programs (Kedharnath and Mathews 1962). As teak takes at least 20 years to come to flowering, the induction of early flowering and fruit set is a clear advantage for breeding purposes. However, for production of wood, flowering and early maturation states are undesirable.

Grafting and budding are the two methods most widely practised in India and the other teak-growing countries. Several different techniques have been used successfully.

2.10.1 Root Grafts

To ensure flowering within 1 or 2 years, grafting of flowering twigs as scions or use of basal buds (the buds at the base of the inflorescence) for budding has been resorted to, with low success rates. Dabral (1977) developed the technique of root grafting to enable a flowering shoot to survive and flower in 1–2 years.

Shoots developing from the basal buds of the previous year were selected for grafting. One-to-two-year-old seedlings were dug out with about 30 cm of the root. The collar diameter of the seedling was of the same thickness as the base of the flowering shoot where the graft was to be made. The root was separated by a cut below the collar and the cut end shaped into a wedge. A notch was cut at the base of the flowering shoot to fit closely with the wedge. The root was then wrapped tightly to the shoot with polyethylene tape and jute thread. A thin metal wire wrapped tightly around the shoot below the graft brings about girdling as the shoot thickens. The graft was covered with a ball of earth (clay:sand:farmyard manure, 2:2:2) and wrapped in jute cloth to facilitate root growth and moisture retention.

The graft establishes in 15 days with the root giving out rootlets, and is ready for transplanting after 3 months, by separation from the mother by excision below the union. May–June was found to be the best season for root grafts in Maharashtra State in the south of India. The advantage of root grafting, as opposed to other techniques in use, is that success rates are high with little loss of buds. Flowering is ensured within 1 or 2 years of grafting. Grafts remain dwarf and form terminal flowers which facilitate easy orchard management. (Dabral 1977)

2.10.2 Rooted Cuttings

Teak is considered to be a difficult-to-root species. Dabral (1977) reported that the methods using hormones adopted by Bhatnagar and Joshi (1972) did not succeed in the areas situated in the South.

Mahmood Husain et al. (1976), however, observed almost 100% rooting in bud sprouts taken from teak stumps and grown under glass house conditions. Buds with two or three pairs of leaves were planted under intermittent mist in polybags containing a mixture of sand, soil, and farmyard manure in equal proportions. Rooting occurred without any hormones and the rooted cuttings could be transplanted to the field in 4–6 months. Such techniques would be of greater importance, if found applicable to woody cuttings from mature trees.

Budding and grafting are now standard procedures followed for establishment of seed orchards in India and in countries like Thailand and Indonesia.

The rates of success obtained with the same techniques differ in different localities. Clones are also known to respond differently to the different grafting methods (Chalmers 1962) and also to the season, the most favorable months being April–June (Rawat and Kedharnath 1968; Dabral 1977). Knud Bryndum (1969) also found this season the most favorable in Thailand, but for administrative reasons the next best season of November–December was preferred.

Results obtained from existing clonal seed orchards around the world have shown that vegetative propagation is feasible for large-scale establishment of seed orchards (Kedharnath 1983; Hedegart 1976). Variation between clones has been observed in several traits, such as susceptibility to fungi, resistance to insects, fiber length, and age at first flowering, indicating that these are under genetic control.

2.11 Micropropagation

Tissue culture studies on teak were undertaken at the National Chemical Laboratory (NCL), Pune, India in 1979 in collaboration with the Forest Development Corporation of Maharashtra, India. Micropropagation methods were standardized using material both from seedlings and mature trees (Gupta et al. 1980). Elite trees over 100 years old from teak forests at Allapalli in the center of India have been multiplied using this technique. The protocols for multiplication of seedlings and mature trees have already been described in detail (Mascarenhas et al. 1987a) and have been summarized below.

Seedling. Teak seeds were first treated for 12 h with KOH to soften the hard coat. The seeds were then separated from the fruits using a nutcracker and healthy seeds sterilized with 0.07% $HgCl_2$ solution for 2 min and then washed with sterile water several times before inoculation on semi-solid White's (1962) medium in test tubes for germination.

Nodal segments were excised from seedlings and treated with a solution of polyvinyl pyrollidone (PVP) and 2% sucrose for 30 min on a shaker. The segments were washed with distilled water and inoculated on a semi-solid modified Murashige and Skoog's (1962) medium containing kinetin, benzyladenine, and sucrose and incubated at 28 °C in light. Three to four shoots developed from each node in 20–25 days. Rooting was induced by transfer of shoots to liquid White's (1962) medium containing a mixture of the auxins IAA, IBA, IPA, and NAA for 48 h in dark. The shoots were then transferred to semi-solid White's medium and further incubated in light.

Mature Trees. Terminal buds (1–2 cm long) were taken from the upper branches of the trees. Buds were surface sterilized with 0.15% $HgCl_2$ for 10 min and then washed with sterile water and inoculated. The medium and incubation conditions were the same as that used for seedlings. Rooting of the shoots required a higher concentration of the auxins. The rooting percentage was however lower than that of seedling shoots.

2.11.1 Field Testing of Micropropagated Teak

Plantlets of both seedling and mature tree origin were hardened by transfer to a sterile mixture of soil, sand, and compost in polybags. Plantlets were ready for transfer to field in 40–60 days, when they attained a height of about 30–40 cm.

Field trials were also conducted on the tissue-culture-raised plantlets from seedlings and mature trees and compared with the growth of seed raised progeny of the elite trees. Tissue-cultured plants showed increased growth rates in height and girth over the control plants up to the 4th year (Mascarenhas et al. 1987a). The increments were more or less even by the 7th year, though the absolute measurements were still higher for the tissue-cultured plants (Mascarenhas et al. 1988a).

All the tissue-culture-raised plants of seedling and mature tree origin flowered in the 3rd year of planting, whereas only a small percentage of seed-raised plants in the experimental plot flowered in the 4th year. In nature, the normal age of flowering of teak is around 20 years. Although age of flowering is reported to be genetically controlled (Larsen 1966), the environment also plays a major role in the expression of this trait. Culture conditions seem to influence this characteristic, as evidenced by the earlier flowering of tissue-cultured plants of seedling origin. While early flowering in the tissue-cultured plants is of advantage in seed orchards, its effect on the timber quality cannot be ascertained at this stage. The increased growth rates in tissue-cultured plants produces stems suitable for poles, much used in construction work in India, earlier than for normal plants (Mascarenhas et al. 1988a). Thinning of 50% of the plantation after 7 to 10 years of planting is the usual practice in teak plantations in Maharashtra (Oka 1976). Tissue-cultured teak could be profitably harvested earlier (Mascarenhas et al. 1988a).

The early flowering characteristic can affect the stem form due to the terminal flowering bud. Hence the use of tissue culture in operational planting will depend on an evaluation of the field trials after a sufficient number of years. The use of tissue culture for quicker multiplication in teak improvement is, however, justified by its application in the quicker establishment of large-scale clonal seed orchards.

Mention must be made here of the possibility of utilizing plantlets obtained through somatic embryos in operational forestry. Such plantlets are presumed to be juvenile in character and hence early flowering may be avoided. Work on this aspect has been carried out by Lakshmisita and Chattopadhayay (1986), who report obtaining the early stages of embryogenesis in callus cultures obtained from shoots and endosperm. Plantlets obtained from endosperm will be triploids and may exhibit improved vigor and growth. The use of in vitro cultures as a method of germplasm conservation is being investigated by Chaturvedi (1983), who has maintained excised root cultures for long periods in liquid media.

3 Tamarind (*Tamarindus indica*; Cesalpinaceae)

Tamarind is an evergreen tree indigeneous to Africa and India. The tree has multiple uses, but the most important product is the tamarind fruit, which is used in commerce.

3.1 Distribution

Tamarind is found cultivated or naturalized in many parts of the tropics and subtropics. In India it is found throughout the plains and sub-Himalayan tracts and is a common shade tree on the road sides. The tree is drought-tolerant and grows on a wide range of soils. Hence it is grown for afforestation on degraded and rocky lands in arid areas.

3.2 Biology

Tamarind is evergreen except in very dry areas, where it remains leafless for a short time in the hot season (Anon. 1976a). The tree has a short, thick trunk and a dense, spreading crown of feathery foliage. The leaves are paripinnate with subsessile, oblong leaflets. The flowers, which appear from April to May, are small and yellowish with pink streaks and occur in racemes. Fruits are pods with shallow constrictions between seeds. The seeds are enveloped by the leathery endocarp, covered with the sour-sweet pulp.

3.3 Diseases and Pests

Tamarind is attacked by rots caused by fungi like *Xylosia euglossa*, *Polyporus calensis*, and *Trametes floccosa*. Fruits and the seeds are attacked by insects not only on the tree but also during storage. The nematodes *Xiphinema citri* and *Longidorus elongatus* are found associated with the roots of tamarind.

3.4 Uses of Fruit Pulp

The fruit pulp of tamarind is a common culinary souring agent in India. Refined pulp and concentrates of tamarind are now exported. The most important constituents of tamarind pulp are tartaric acid (8–10%) and invert sugars (30–40%).

3.5 Uses of Seed

Seeds of tamarind are the source of tamarind kernel powder (TKP), which finds use as a sizing material in the textile industry. TKP consists of up to 60% of a polysaccharide called jellose, which is an excellent substitute for fruit pectin, as it forms a strong gel. Other uses of TKP are as an adhesive and as a creaming agent for concentration of rubber latex.

3.6 Timber

Tamarind also produces a strong, hard, and heavy timber used for agricultural implements, tools, furniture, and construction work. The wood is also a very good fuel.

3.7 Varieties

Two varieties of tamarind are known, the brown and red pulped. The latter variety has pulp that is red-coloured due to a pigment – chrysanthemin, and is sweeter than the brown-pulped variety. Though genetic and provenance studies have not been carried out, tamarind trees with yields up to five times higher than average and with bigger pods have been identified.

3.8 Propagation

Tamarind is easily propagated by seed but does not breed true for yield and fruit pulp characteristics. Vegetative propagation in tamarind has, however, not been widely practised in the past. Techniques like air-layering, inarch grafting, and micropropagation have been developed only recently.

3.8.1 Vegetative Propagation

Air Layering. Vegetative propagation of tamarind was studied by Bhore (1985) in the red-pulped and brown-pulped varieties. The procedure involved removing about 4 cm of bark from the stem 4–5 cm below the apical end. Lanolin paste with IBA at various concentrations was applied to the cut and covered with wet sphagnum moss enveloped in a polythene sheet. Air layers were separated from mother plants in 3–4 months and percentage of success recorded after 6 months. A 50 and 60% success was achieved using 4000 ppm IBA for the red and brown varieties respectively. Controls without IBA gave 10 and 18% response and 5000 ppm gave 40 and 54% success respectively.

Inarch Grafting. Grafting of the brown variety of tamarind was also successfully carried out by Bhore et al. (1985) using 2-year-old root stocks with a stem thickness of 0.7 to 0.8 cm. Grafts were detached after 5 months and 95% success recorded after 6 months of planting.

Clonal propagation of tamarind trees showing desirable fruit pulp characteristics thus seems possible using either of the two above-mentioned methods, although further progress on these methods has not been reported.

3.8.2 Micropropagation of Tamarind

Tissue culture studies on tamarind were initiated at NCL in 1978 with the aim of large-scale propagation of elite trees having five times the average fruit yield. Methods of propagation using nodal explants from seedlings and mature trees were developed in 1981 (Kulkarni et al. 1981; Mascarenhas et al. 1981). The protocols for plantlet production using seedling nodal explants have been published earlier (Mascarenhas et al. 1987b) and are summarized below. The techniques, however, still require optimization for mature tree explants.

Seedling. Seeds of tamarind were surface sterilized and germinated on a simple solid medium. Explants 8–10 mm long containing one node were inoculated on a medium consisting of the minerals and vitamins of Murashige and Skoog (1962) supplemented with kinetin, benzyladenine, calcium pantothenate, and biotin. Regenerated shoots were subcultured on the same medium every 6 weeks.

Rooting of shoots was accomplished by transfer of 40–50-mm-long shoots to a liquid White's (1962) medium containing a mixture of the auxins IAA, IAP, NAA on filter paper supports for 72 h in dark, followed by transfer to the same medium without hormones. Root initiation took place in 2 weeks. Plantlets were then transferred to a sterile mixture of sand, soil, and compost in polybags. Plantlets were hardened in a greenhouse until they attained a height of 300–350 mm. Survival of plantlets was 50–60% (Mascarenhas et al. 1981).

Mature Trees. Shoot cultures were isolated from nodal explants cultured on the same medium used for seedlings. Rooting of these shoots, however, required a higher concentration of the auxins and further transfer to solidified half-strength Murashige and Skoog medium containing activated charcoal for elongation of roots. Rooting percentage was very low, being below 20% (Mascarenhas et al. 1981).

3.8.3 Field Trials on Micropropagated Tamarind

Plantlets obtained from seedling cultures have been growing in the field for 8 years. The most striking observation in the field trials was the occurrence of the flowering and fruiting in 3.5 years of planting (Mascarenhas et al. 1987b). It is a general observation that grafts made from mature trees come to flower in a shorter period. In this case, however, the plantlets were of seedling origin. Normally, tamarind trees take 10–12 years to flower (Anon. 1976b). Micropropagated plants also showed profuse branching and a better spread than the controls. This fact and the smaller size of the micropropagated trees would simplify orchard management and harvesting operations. A higher planting density of 156 plants per hectare is possible at 9 × 9 m spacing. These gains can more than offset the higher cost of micropropagated plants. The use of these trees for timber still remains to be ascertained, although grafts are generally unsuitable for this purpose.

4 Bamboo (Gramineae)

The bamboos are an important component of the forests of Asia and South America. So indispensable is bamboo in many Eastern nations that it permeates the folklore and myths. The versatility of bamboo is not only restricted to rural and household uses, such as for construction and manufacture of implements, but today it is also a major raw material for the paper industry (Varmah and Pant 1981). Bamboos are suitable for plantations as they can be cultivated at short rotations and are fast-growing and adaptable to a wide range of climates.

The inclusion of bamboos in this chapter on hardwoods is justified if hardwoods are defined as angiosperms and softwoods as conifers. From a utilitarian point of view, bamboos serve much the same purposes as other trees.

4.1 Biology

Bamboos are perennial arborescent grasses characterized by woody underground rhizomes from which the aerial stems, called culms, arise. All the three species discussed in this chapter viz. *Bambusa arundinacea*, *B. vulgaris*, and *Dendrocalamus strictus*, have a short and thick rhizome termed pachymorph (McClure 1966), and occur in dense clumps. Culms are smooth, rounded, and jointed. The arrangement of fiber bundles, the thickness of the outer region, and the deposit of silica make them very hard (Anon. 1948). New culms sprout at the beginning of the rains and grow very rapidly at a rate of about 8 cm per day and achieve their full size in about a month.

Bamboos exhibit a flowering behavior that is as yet unexplained. Many bamboos flower once in a lifetime, which may range from 12 to 120 years. Entire populations show synchronized flowering and seeding and then die out. Availability of seeds for artificial plantations is thus a problem.

4.2 Uses

Bamboo culms are suitable for wide variety of purposes, having qualities like lightness combined with hardness and strength. The smooth, straight, and hollow stems can be split with ease. A variety of functional objects for daily use are made out of bamboo (Varmah and Pant 1981; Anon. 1976c). It is particularly useful as a cheap timber substitute for the construction of houses.

4.3 Species

Bamboos consist of about 550 species belonging to 30 genera. In this section only three species have been considered viz. *Bambusa arundinacea*. *B. vulgaris*, and *Dendrocalamus strictus*. These are the most important species grown in India for which methods of vegetative propagation have been developed.

4.3.1 *Bambusa arundinacea* Wiild.

This species forms large, dense clumps of green thorny culms that attain a height of 25 to 30 m.

B. arundinacea grows wild throughout India and is also found in Burma and Sri Lanka. It prefers rich moist soils and grows best in moist deciduous forests where the annual rain fall is 2000–3000 mm. The culms are thick-walled and yield good quality paper pulp but extraction and utilization is rendered difficult due to its thorny and crooked nature. It flowers generally once in about 30 years and then dies out. *B. arundinacea* can be selfed and seeds do not have dormancy (Kondas et al. 1973).

Kondas et al. (1973) identified four types of seedlings in *B. arundinacea* viz. grassy, grassy erect, erect, and very erect. The last two types are more vigorous and attain harvestable size earlier. The trait seems to be under genetic control. Kondas (1982) also reported some varieties with culms free of thorns and branches. Vegetative propagation of such varieties will be useful in improvement of plantations.

4.3.2 *Bambusa vulgaris* Schr.

This is the most widely cultivated bamboo species around the world. Vegetative propagation is relatively easy and flowering is rare.

B. vulgaris is cultivated throughout the tropics but its origin is unknown. Several varieties of this bamboo can be distinguished based on the color forms. It is adaptable to a wide range of soils and moisture conditions, but is sensitive to cold.

B. vulgaris is known to flower only very rarely and that, too, restricted to a few plants. Hence the populations of this species have apparently remained always in a vegetative state. Culms of this species are very strong and find use of props and also in the paper industry.

4.3.3 *Dendrocalamus strictus (Roxb.) Nees*

This is the commonest and most widely distributed of all Indian bamboos. Being hardy and tolerant to drought and frost, it is suitable for plantations in areas where the other species of bamboo do not thrive. This species forms an important source of paper pulp in India. It is used as props, as the culms are very strong. *D. strictus* also shows variation in growth form, but whether the variations are genotypic is not clear (McClure 1966). *D. strictus* has a flowering cycle of 25–35 years but sporadic flowering has been noticed annually (McClure 1966).

4.4 Vegetative Propagation

The various methods of vegetative propagation used for different bamboos have been described by Troup (1921) and McClure (1966).

Division of Rhizomes. The method consists of dividing the mass of rhizomes and transplanting the culms with two or three rhizome sections attached. *B. arundinacea* is often propagated using this method. Offsets are cuttings of the culm with the rhizome. One-year-old culms are cut about 90–120 cm from the ground and their rhizomes dug up with the roots intact and cut off to a length sufficient to include a well-developed bud. Offsets are planted deep enough to cover the first two to three nodes.

Rhizome Cuttings. Sections 15–30 cm long of the rhizomes of the previous year, containing at least one bud, are planted.

Layering. Culms are bent over and pegged down below the ground. When roots and shoots develop at the nodes, the internodes are cut and the plants separated. *B. vulgaris* is easily propagated in this way.

Air Layering. Abeels (1962) described a method wherein a funnel-shaped contrivance is made of the bamboo sheath around a node and filled with soil and moss. Healthy and leafy twigs root quickly at the node if the stem was injured. These can then be separated and planted in the nursery.

Stem Cuttings. Abeels (1962) described methods for raising rooted stem cuttings using stem segments with one to several internodes. In the first method cuttings were directly planted in soil in the nursery. In a variation of this method which gave good results, water was filled in the hollow internodes through a hole in the middle. The upper cut ends were coated with a plastic compound to prevent loss of moisture. *B. vulgaris* responded well, producing shoots in 15 days and roots in the 4th week.

In another method, entire branches cut at the furthest node were suspended from a wooden lattice supported on drum floats, in a pond with up to 10 cm of the base of the cuttings immersed in water. *B. vulgaris* cuttings produced shoots and roots in 5–6 weeks. *D. strictus* failed to produce roots.

Cuttings similar to those mentioned above planted in the sandy bed of a stream with 5–10 cm of the base buried gave about 100% rooting in *B. vulgaris*. Transplanting was also done successfully. Varmah and Pant (1981) reported that two node stem sections of *D. strictus* gave good results in April. Horizontal planting was found superior to the vertical or oblique positions. Culm cuttings were successful with *B. vulgaris* when the middle one third of the culm consisting of two nodes was used where culms could be harvested 2 years after planting (Abeels 1962).

4.5 Tissue Culture of Bamboos

The various methods of vegetative propagation described above are applicable only on a small scale, as the multiplication rates are low. Many of the methods involve use of propagules that are quite bulky to handle or transport.

Tissue-culture methods offer an alternative for rapid and large-scale vegetative propagation of bamboos (Kondas 1982; Dogra 1981). Besides the high multiplication rates, plantlets comparable in size to seedlings are possible by tissue culture, thus simplifying plantation establishment.

Propagation of bamboos by tissue culture has been successfully carried out by two different methods. Plantlets have been regenerated through somatic embryogenesis from callus initiated from seed of *B. arundinacea* (Mehta et al. 1982), *D. strictus* (Rao et al. 1985) and *Sinocalamus latiflora* (Yeh and Chang 1987). Callus obtained from young inflorescences have similarly regenerated plantlets in *B. oldhamii* (Yeh and Chang 1986a) and *B. beecheyana* (Yeh and Chang 1986b).

Rapid multiplication of bamboos is also possible through micropropagation using nodal explants from seedlings and mature culms (Nadgir et al. 1984). In *D. strictus*, over 10000 plantlets could be obtained in 1 year starting with one seedling. Culture in a liquid medium in shake flasks gave better results than solid media. Eight to ten shoots developed in 6–7 weeks of culture, and this multiplication rate was maintained in subsequent subcultures. Spontaneous rooting was observed in 40% of the shoots growing in the flasks. Of the remaining shoots 80% could be rooted after treatment with 0.1 mg/l IBA for 48 h. When plantlets attained a height of 50–60 mm they were transferred to a sterile mixture of sand and soil (1 : 1) in pots and hardened off. When transferred to the field 70–80% of plantlets survived.

Culture of nodal segments from mature culms of these species of bamboo, viz, *D. strictus*, *B. vulgaris*, and *B. arundinacea*, was also attempted by Nadgir et al. (1984). In all the species, shoot proliferation occurred on solid media in 3–4 weeks in 80% of the cultures. Two to three shoots developed from a node. Liquid shake flask cultures were not beneficial as in the case of seedling nodes. The shoots obtained from mature explants, however, did not root with any of the treatments tried except in *D. strictus*, where 20% of the shoots could be rooted if treated with 1.0 mg/l IBA for 96 h in the dark followed by transfer to a solid medium containing 0.25% activated charcoal.

4.5.1 Field Testing of Micropropagated Bamboo

Tissue-culture-raised plantlets from seedling explants of *D. strictus* were planted out for pilot scale trials (Mascarenhas 1987; Mascarenhas et al. 1988b). Culm formation in tissue-cultured plants took place in 30 months as compared to seedlings, which required 48 months. After 52 months, the height of the main culm, the number of culms, the number of internodes in the main culm, and the girth of the second internode nearly doubled, as compared to those in the controls (Mascarenhas et al. 1988b). Thus culms of harvestable size were formed much earlier in the micropropagated bamboo than in seedlings. These results, though observed in small-scale trials, give an indication of the potential for in vitro multiplication of bamboo seedlings by reducing the rotation cycles. Propagules can also be produced in large numbers throughout the year. Micropropagation of elite mature culms, even if successful, carries with it the possibility of simultaneous flowering and dying out of whole plantations with the original mother clumps. Plantlets of seedling origin will probably grow till the normal flowering age.

Propagation of bamboos can now be considered on a much larger scale than is possible by conventional vegetative propagation methods. The use of in vitro techniques, though unsuccessful at present with mature bamboos, is justified in the case of seedlings, because of the benefits mentioned earlier, that occur due to the shortening of the rotation period. In vitro techniques can be used for multiplication when seed supplies are scarce or when hybridization between species are attempted.

4.5.2 In Vitro Flowering of Bamboos

The peculiar flowering behavior of bamboos has been a limitation in the application of breeding techniques for improvement. Because of its unpredictable nature and occurrence at long intervals, the phenomenon has not been amenable to study. In a major breakthrough in the study of bamboos, induction of flowering in tissue cultures of three species of bamboo was achieved (Nadgauda et al. 1990). The implications of the reported results are tremendous. Not only can the flowering phenomenon now be studied in all its aspects in the laboratory, but it also raises the possibility of seed production in culture vessels throughout the year. Hybridization between species can now be attempted in the laboratory.

5 Conclusions

The various vegetative propagation methods described in this chapter were developed primarily to solve problems faced in traditional breeding and large-scale propagation of the tree species. Success has been obtained in varying degrees with the three trees, and this leads to the question of the relative usefulness and cost effectiveness of the different methods. Conclusions, however, have to be tentative at this stage, for many of the methods for which field trials are in progress and would require several years before any evaluation could be made with confidence. Graft incompatibility, the flowering age in bamboos, the effect of early flowering on timber characteristics, etc. are some of the factors which will become manifest only after a certain period of time. In the establishment of clonal seed orchards in teak, stump budding and root grafts appear promising due to higher percentage of success over other conventional methods. The number of buds that can be safely taken from an elite tree without affecting its growth is limited and establishment of seed orchards on a large scale is thus hampered. Micropropagation on the other hand can produce clones in much larger numbers from fewer buds. In vitro techniques can also ensure pathogen-free trees in the seed orchards. All three techniques, however, require skilled workers and are labor-intensive. Micropropagation, particularly, is an expensive procedure. In tamarind, air layering and inarch grafting have been successful, but this involves utilizing numerous twigs per tree. Micropropagation can again produce large numbers of plants starting from a few buds. The conventional vegetative propagation in bamboo is highly laborious as it involves bulky propagules. Moreover, multiplication rates are extremely low to meet plantation requirements. Micropropagation, on the other hand, offers a better method for establishing bamboo plantations as high multiplication rates and seedling-like propagules are obtained. Propagation by rooted cuttings and micropropagation have the greatest potential for production of planting stock. To be cost-effective the methods have to be optimized. Success with cuttings can be expected to be achieved in an increased number of tropical species, when modern horticultural facilities are more widely available in the tropics. In regions where cuttings can be produced cheaply in sufficient numbers, micropropagation with its present procedures would be uneconomical.

6 Future Prospects

Relatively small-scale trials such as those mentioned have shown clear differences between in vitro raised plants and those raised by other methods. Our understanding of the reasons for such differences is, however, inadequate. The responses like early flowering, early culm formation, and quicker growth rates in tissue-culture plants appear at first glance to be an effect of the culture conditions rather than genetical or physiological, as seedlings and mature trees respond alike. Conventional vegetative propagation shows no such parallel response. Progeny testing of the tissue-cultured plants could lead to a better understanding of the nature of the response.

The most important factor that limits the use of tissue-cultured plantlets in forestry is its higher cost of production as compared to conventional seedlings or rooted cuttings. Development of mechanized procedures to replace the highly labor-intensive stages in plantlet production can bring down labor costs, which forms the major part of the total cost of the plantlet. The use of robots for handling shoot cultures is also being studied for this purpose (de Bry 1986).

Somatic embryogenesis, which offers much by way of cutting down costs of large-scale propagation, has still not been successful in several of the hardwoods including teak and tamarind, for which micropropagation has been standardized. Somatic embryos, if induced in cell suspensions, can be cultured in bioreactors in very large numbers, thus reducing space and labor requirements. When artificial seeds produced by encapsulating embryos in a suitable matrix become a reality, the benefits of somatic embryogenesis can be further extended. Artificial seeds form an efficient method of plant delivery similar to natural seeds.

Plantlets obtained through somatic embryos will have to be carefully screened in the field after a sufficient number of years to ascertain the growth performance and maintenance of clonal fidelity. This is the case with all in vitro regeneration systems, particularly those involving callus cultures.

All points considered, micropropagation will play an important role in overcoming some of the hurdles in tree improvement. Long-term improvement will have to depend on traditional breeding approaches. Genetic engineering by recombinant DNA techniques is expected to be widely applicable to trees in the near future. Regeneration of plants from transformed cells and tissues and their multiplication to sufficient numbers will require development of in vitro techniques. The contribution of tissue culture to clonal forestry and for the improvement of forests is thus quite promising.

References

Abeels P (1962) Multiplication of bamboos. Indian For 88:481–487

Anon (1948) Bamboos In: The wealth of India, vol 1 (raw materials). Manjunath BL (ed) CSIR, New Delhi, p 145–154

Anon (1976a) Tamarind. In: The wealth of India, vol 10 (Raw materials). Bhadha YR (ed) CSIR, New Delhi, p 114–122

Anon (1976b) Teak. In: The wealth of India, vol 10 (raw materials). Bhadha YR (ed) CSIR, New Delhi, p 136–150

Bawa KS (1976) Breeding of tropical hardwoods: an evaluation of underlying bases, currrent status and future prospects. In: Burley J, Styles BT (eds) Tropical trees: variation, breeding and conservation. Academic Press, New York London, pp 43–59

Bhore DP (1985) Studies on vegetative propagation of tamarind (*Tamarindus indica*) Maharashtra J Hortic 3 (1): 55–58

Brazier JD, Hughes JF, Tabb CB (1976) Exploitation of natural tropical forest resources and the need for genetic and ecological conservation. In: Burley J, Styles BT (eds) Tropical trees: variation, breeding and conservation. Academic Press, New York London, pp 1–10

Bryndums K (1969) Budding of Teak, Indian For. 95:155–157

Campinhos E, Ikemori YK (1980) Mass production of *Eucalyptus* spp. by rooting cuttings. In: IUFRO Symp Genetic improvement and productivity of fast growing trees, Sao Pedro, Sao Paulo, Braz.

Chalmers WS (1962) The breeding of pine (*Pinus caribaea*) and teak (*Tectona grandis*) in Trinidad, some early observations. Caribb For 23:100–111

Chaturvedi HC (1983) Clonal propagation through tissue culture – some critical aspects of clonal propagation and germplasm preservation of plants through tissue culture. In: Proc Natl Symp Advancing frontiers of plant science. Jodhpur, India, pp 120–121

Dabral SL (1976) Extraction of teak seeds from fruits, their storage and germination. Indian For 102:650–658

Dabral SL (1977) Propagation of teak by root grafts. Indian For 103:225–230

De Bry L (1986) Robots in plant tissue culture: an insight, IAPTC Newslett 49:2–22

Dogra PD (1981) Forest genetics – research and application in Indian forestry II. Indian For 107:263–288

Evans J (1982) Plantation forestry in the tropics. Univ Press, Oxford

Gupta PK, Nadgir AL, Mascarenhas AF, Jagannathan V (1980) Tissue culture of forest trees – clonal multiplication of *Tectona grandis* (Teak) by tissue culture. Plant Sci Lett 17:259–268

Hedegart T (1976) Breeding systems, variation and genetic improvement of teak (*Tectona grandis*) In: Burley J, Styles BT (eds) Tropical trees: variation, breeding and conservation, Academic Press, New York London, pp 109–121

Kedharnath S (1983) Genetics and forest tree breeding. In: Jaiswal PL (ed) Genetical research in India. ICAR, New Delhi, pp 181–190

Kedharnath S (1984) Forest tree improvement in India. Proc Indian Acad Sci Plant Sci 93 (3):401–412

Kedharnath S, Mathews JD (1962) Improvement of teak by selection and breeding. Indian For 88:277–284

Kondas S (1982) Biology of two Indian bamboos, their culm potential and problems of cultivation. Indian For 108:179–188

Kondas S, Sree Rangaswamy SR, Jambulingam R (1973) Performance of *Bambusa arundinacea* seedlings in nursery. Madras Agric J 60(9–12):1719–1726

Kulkarni VM, Gupta PK, Mehta UJ, Mascarenhas AF (1981) Tissue culture of woody trees. Clonal propagation of *Tamarindus indica* by tissue culture. In: Proc 6th All India Plant tissue culture Conf, Poona Univ, pp 34–35

Kumaravelu G (1979) Clonal identification of *Tectona grandis* by isozyme studies. Indian For 105:716–719

Lakshmisita G, Chattopadhayay S (1986) Improvement of Forest trees by tissue culture. In: Reddy GM (ed) Plant, cell and tissue culture of economically important plants. Proc Nat Symp, Hyderabad, India, p 195

Larsen (1966) Genetics in teak (*Tectona grandis*). In: R Vet Agric Coll Year Book, Copenhagen

Leakey RRB (1987) Clonal forestry in the tropics: a review of developments, strategies and opportunities. Commonw For Rev 66(1):61–75

Mahmood Husain AM, Somasundarm TR, Subramanian KN (1976) A recent advance in teak culture. Indian For 102:531–532

Mascarenhas AF (1987) Studies on tissue culture raised trees. In: Proc Sem Tissue culture of forest species, Kuala Lumpur, June 15–18 pp 69–75

Mascarenhas AF, Gupta PK, Kulkarni VM, Mehta UJ, Iyer RS, Khuspe SS, Jagannathan V (1981) Propagation of trees by tissue culture. In: Rao AN (ed) Tissue culture of economically important plants. COSTED, Singapore, p 175

Mascarenhas AF, Kendurkar SV, Gupta PK, Khuspe SS, Agrawal DC (1987a) Teak. In: Bonga JM, Durzan DJ (eds), Cell and tissue culture in forestry, vol 3. Nijhoff, Dordrecht, pp 300–315

Mascarenhas AF, Nair S, Kulkarni VM, Agrawal DC, Khuspe SS, Mehta UJ (1987b) Tamarind. In: Bonga JM, Durzan DJ (eds) Cell and tissue culture in forestry, vol 3. Nijhoff, Dordrecht, pp 316–325

Mascarenhas AF, Khuspe SS, Nadgauda RS, Gupta PK, Khan BM (1988a) Potential of cell culture in plantation forestry programmes. In: Hannover JW, Keathley DE (eds) Genetic manipulation of woody plants. Plenum, New York, p 391

Mascarenhas AF, Khuspe SS, Nadgauda RS, Gupta PK, Muralidharan EM, Khan BM (1988b) Biotechnological application of plant tissue culture to forest species in India. In: Proc Int Worksh Application of biotechnology in forestry and horticulture. TERI, New Delhi, Jan 1988 pp 73–77

McClure FA (1966) The bamboos: a fresh perspective. Harvard Univ Press

Mehta UI, Rao VR, Mohan Ram HY (1982) Somatic embryogenesis in bamboo In: Fujiwara A (ed) Proc 5th Int Congr Plant tissue and cell culture, Tokyo, p 109

Murashige T, Skoog F (1962) A revised medium for rapid growth and bioassays with tobacco tissue cultures. Physiol Plant 15:473–497

Nadgauda RS, Parashurami VA, Mascarenhas AF (1990) Precocious flowering and seeding behavior in tissue cultured bamboo. Nature (London) 344:335–336

Nadgir AL, Phadke CH, Gupta PK, Parasharami VA, Nair S, Mascarenhas AF (1984) Rapid multiplication of bamboo by tissue culture. Silv Genet 33:219–223

Nanda KK (1962) Some observations as growth, branching behaviour and flowering of teak (*Tectona grandis*) in relation to light. Indian For 88:207–218

Oka AF (1976) Teak nursery practices in Maharashtra. Res Educ Circle, Nagpur, India

Rao IO, Rao IVR, Narang V (1985) Somatic embryogenesis and regeneration of plants in the bamboo *Dendrocalamus strictus*. Plant Cell Rep 4:191–194

Rawat MS, Kedharnath S (1968) Field grafting and budding in teak (*Tectona grandis*). Indian For 94:260–261

Seth SK, Khan MAW (1958) Regeneration of teak forests. Indian For 84:455–466

Troup RS (1921) The silviculture of Indian trees, vol 2. Clarendon, Oxford

Varmah JC, Pant MM (1981) Production and utilization of bamboos. Indian For 107:465–476

White PR (1962) The cultivation of animal and plant cells. Ronald, New York

Yeh ML, Chang WC (1986a) Plant regeneration through somatic embryogenesis in callus culture of green bamboo (*Bambusa oldamii* Munro). Theor Appl Genet 73:161–163

Yeh ML, Chang WC (1986b) Somatic embryogenesis and subsequent plant regeneration for inflorescence callus of *Bambusa beecheyana* Munro var. *Beecheyana*. Plant Cell Rep 5:409–411

Yeh ML, Chang WC (1987) Plant regeneration via somatic embryogenesis in mature embryo-derived callus culture of *Sinocalamus latiflora* (Munro). Plant Sci 51:93–96

Zobel B, Talbert J (1984) Applied forest tree improvement. John Wiley & Sons, New York, pp 310–328

Zobel B, Campinhos E, Ikemori Y (1982) Selecting and breeding for wood uniformity. In: Proc TAPPI Res Dev Meet, Ashville, NC, pp 159–168

11 Clonal Forestry with Yellow-Cedar

J.H. Russell[1]

1 Introduction

Stecklings have been the major propagule type for the reforestation of yellow-cedar [*Chamaecyparis nootkatensis* (D. Don) Spach] over the past 15 years in British Columbia. Over 70% of the planting stock since 1975 has been stecklings, and currently, one-half to three-quarters of a million stecklings are planted annually.

Clonal forestry, however, is in its infancy for yellow-cedar. The original objective of the steckling program was to vegetatively multiply seeds in short supply (Karlsson 1981; Russell 1988). To date, no tested, genetically improved clones have been used for reforestation. However, curently over 1000 clones are in tests, and the first roguing of a clonal cutting-orchard will occur in 1992, based on 5-year data. This will result in the deployment of selected clones for reforestation by 1994.

The purpose of this chapter is to discuss the development of yellow-cedar clonal forestry in British Columbia from its beginnings to the present-day program to future strategies. Emphasis will be placed on the influence of historical, economical, and biological factors and events that have shaped the present-day program, and that will influence future strategies.

2 Background

2.1 Natural Range and Autecology

Yellow-cedar occurs from Northern California to Alaska. In Oregon and Washington, it occurs on the west side of the Cascade Mountain Range and on the Olympic Penninsula. From southern British Columbia to Alaska, yellow-cedar occurs in a narrow band along the coast. There are two isolated inland populations, one east of the Cascades in northern Oregon, and the other east of the Monashee Mountains near Slocan Lake in British Columbia.

[1] British Columbia Ministry of Forests, Cowichan Lake Research Station, Box 335, Mesachie Lake, British Columbia, Canada, VOR 2N0

Clonal Forestry II, Conservation and Application
Ed. by M.R. Ahuja and W.J. Libby

Yellow-cedar typically occurs in cool, humid environments with relatively short growing seasons. Winter temperatures are not too severe, and snowfall is heavy (Andersen 1959). The species occurs on wet, nutrient-poor to nutrient-rich sites throughout its range, and is outcompeted on well-drained, deep soils although it has the potential to perform well on these sites (Andersen 1959). Yellow-cedar occurs at high elevations in the southern part of its range, and from sea level to timberline north of Knights Inlet in British Columbia. It grows mainly in mixtures with other species including western redcedar (*Thuja plicata*), western hemlock (*Tsuga heterophylla*), mountain hemlock (*Tsuga mertensiana*), sitka spruce (*Picea sitchensis*), and subalpine fir (*Abies lasiocarpa*).

2.2 Supply and Demand

Yellow-cedar, with its typically light-colored wood and uniform, tight grain, is a valuable and highly sought-after timber species. Although a minor component of British Columbia's annual timber harvest, the best-grade logs have the highest economic value of any species.

An increase in high-elevation logging in the mid-1970s coupled with a favorable export market resulted in an increase in harvesting of yellow-cedar, and a subsequent demand for planting stock for reforestation. This demand could not be met by conventional nursery-grown seedlings due to infrequent, consistently small cone crops, and poor and erratic germination. Thus, in order to meet the increase in demand for planting stock, which could not be supplied by seedlings, a system for the large-scale production of stecklings was developed (Karlsson 1981).

3 Development of the Steckling Program

3.1 Early Development

In the early 1970s, initial nursery rooting trials using young cutting-donors were highly successful with respect to rooting and steckling form and growth (Karlsson 1974). Subsequent research trials concentrated on improving rooting and quality of planting stock through cultural manipulations, and on investigating the effects of age of cutting-donor on rooting. A significant breakthrough came with the development of a system that produced a plantable steckling in the nursery in one growing season (Karlsson 1974).

These initial successes prompted the continuation and expansion of the yellow-cedar steckling program. From 1975 to 1980, the research focus was on developing a large-scale production system for rooting cuttings in the nursery. A system was developed that was tailored towards existing nursery technology, and that was biologically successful and cost-efficient (Karlsson 1981).

Early research on maturation effects showed the need for hedging of cutting-donors in order to maintain juvenility with respect to rooting success and steckling performance. This also allowed for the repeated use of each cutting-donor for a number of years, an important factor given the limited seed supply. A 1-ha cutting-orchard was established at Cowichan Lake Research Station on Vancouver Island in 1978/79 comprising 27 open-pollinated families, the seed of which originated from phenotypically selected parents (Fig. 1).

Fig. 1. Yellow-cedar cutting-orchard at Cowichan Lake Research Station. This orchard was established in 1978/79 with 2-year-old seedlings from open-pollinated families and has been hedged annually to 25 cm since 1985. The orchard contains over 5000 plants and is capable of producing 1/2 million cuttings annually

Fig. 2. a Recently rooted yellow-cedar cuttings, **b** a crop of 300 000 rooted cuttings in the greenhouse

Large-scale propagation of stecklings for reforestation began in the late 1970's concurrent with the acceptance of stecklings by nurserymen and field foresters as a viable alternative propagule to seedlings (Fig. 2). In the mid-1980s operational nursery production of stecklings was transferred to the private sector.

Research since the mid-1980s has focused on (1) increased understanding of cutting-donor maturation effects on rooting and steckling quality; (2) the physiological response of yellow-cedar seedlings and stecklings to environmental stress; and (3) the development of a clonal-forestry program.

3.2 Present Program

As mentioned above, up to 750000 yellow-cedar stecklings are produced annually for reforestation in British Columbia, which satisfies nearly all of the current annual demand. The process is still mainly vegetative multiplication of untested open-pollinated families and seedlots. Several kinds of hedged cutting-donors are used, including container-grown and soil-grown donors from 1 to 15 years in chronological age (Russell et al, 1990). Currently there are four soil-based cutting-orchards, and three container cutting-orchards in operation.

4 Cutting-Donor Management

Rooting, other elements of cutting performance in the nursery, and quality of outplanted stecklings, vary depending on cutting-donor maturation level. Figure 3, compiled from the results of research and operational rooting trials, illustrates the relationship between rooting success and cutting-donor maturity. Rooting success declines with increasing chronological age of the seedling cutting-donor such that four overlapping and arbitrary stages of increasing maturation can be delineated: I. Juvenile (greater than 85% rooting); II. decreasing juvenility ($< 85\% > 75\%$ rooting); III. increasing maturity ($< 75\% > 40\%$ rooting); and IV. mature (less than 40% rooting). Similar relationships with cutting-donor maturation have been shown using rooting traits such as number of main roots or symmetry of main roots, and with steckling traits such as stem straightness or root-collar diameter.

As can be seen from Fig. 3, hedging can delay cutting-donor maturation and improve rooting success and stock quality. For example, cuttings from hedged 5-year-old seedlings have similar rooting success to cuttings taken from 3-year-old unhedged seedlings. Cuttings from hedged 13-year-old seedlings behave similarly to cuttings from unhedged 5-year-old seedlings. Serial propagation is very successful in delaying maturation (Fig. 3). Cutting-donors that have gone through up to four cutting cycles (1 to 2 years between cycles) starting with a 2-year-old seedling (i.e., 5 to 7 years from seed), produce cuttings that root with greater than 90% success. Research in serial propagation has not yet progressed past this stage.

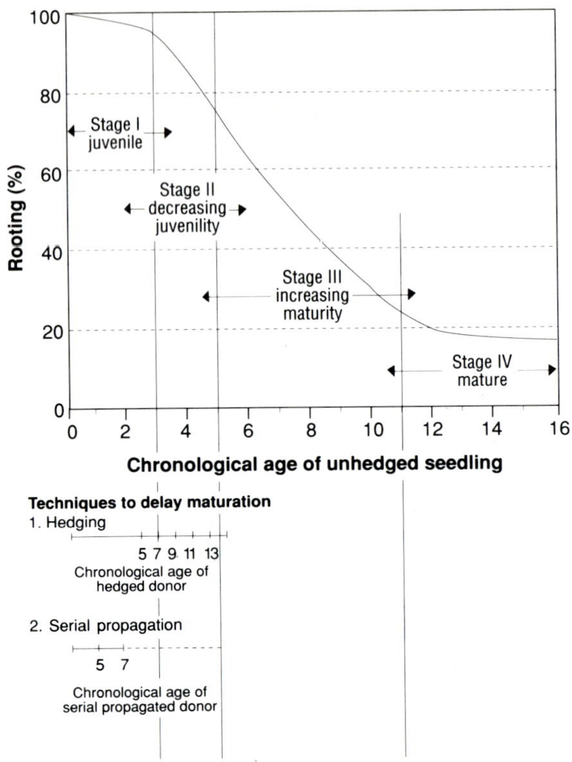

Fig. 3. Yellow-cedar maturation. Four stages of increasing maturity can be identified with respect to rooting success. See text for details. Apparent delays in maturation of hedged and serially propagated donors appears below the seedling-donor graph. (Russell et al. 1990)

Hedging cutting-donors also improves steckling quality. Table 1 presents the effects of hedging and hedging height on steckling traits. All observed traits have been significantly improved if the cutting material originates from donors hedged to 25 and 50 cm.

Donors from maturation stages I and II produce acceptable cutting material for reforestation. These cuttings root with greater than 75% success and the resultant stecklings are similar to seedlings with respect to growth and form. Presently, growers prefer stage-I cutting material, since these cuttings have greater rooting success, lower cull rate, and are easier to root and grow than are cuttings from stage-II donors.

Figure 4 shows that the rate at which rooting occurs, the physiological responses of the cuttings to their environmental conditions, and rooting success are all related to cutting-donor maturation (Grossnickle and Russell 1989). Cuttings from 5- and 12-year-old hedged donors required 12 and 20 weeks, respectively, in the rooting environment before being ready for transfer to a growing environment. Cuttings from older free-growing trees required more time in the rooting environment. The more juvenile the cutting-donors, the less

Table 1. Effect of hedge height of 10-year-old cutting-donors on rooting and subsequent growth and form of 1-year-old yellow-cedar stecklings. (Russell 1988)

Trait	Hedge height (cm)			
	25	50	100	No hedging[c]
% Rooting	81% a[b]	70% b	69% b	44% c
Stem straightness[a]	1.24 a	1.35 a	1.91 b	2.61 c
Root-collar diameter (mm)	3.88 a	3.77 ab	3.41 b	2.98 c
Shoot dry weight (g)	2.56 a	2.19 b	1.72 c	1.31 d
Root dry-weight (g)	0.89 a	0.72 b	0.56 c	0.42 c

[a] Stem straightness, 1 = upright; 2 = 45° to upright; 3 = 0° to 45° from horizontal.
[b] Different letters in the same row indicates significant differences at the P = 0.05 level according to Duncan's multiple range test.
[c] Cuttings were taken from the upper third of the crown on unhedged donors.

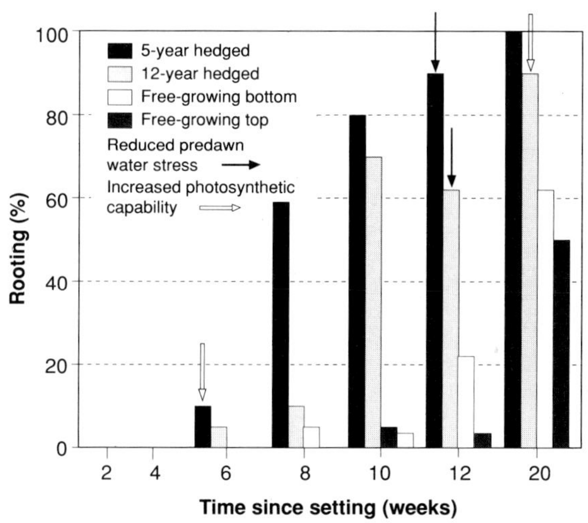

Fig. 4. Rooting percentage of yellow-cedar cuttings over time taken from donors of varying maturity. Times when predawn water stress was reduced, and photosynthetic capability were increased in cuttings, from different donor plants, to levels comparable to those in seedling controls are identified with *arrows*. (Grossnickle and Russell 1989)

physiological stress the cuttings experience, and the more quickly they root and begin to function similarly to seedlings. Significant interactions of cutting-donor maturation and genotype, and nursery cultural regime also occur (Russell et al. 1990). Thus, stage-II cuttings require more attention, are more prone to failure, and cost more to produce a steckling than stage-I cuttings.

Currently most of the cuttings for the operational steckling program originate from donors that are less than 5-years-old. The donors are grown in containers in greenhouses and intensively managed, i.e., pruning, adequate

Fig. 5. **a** A 2-year old yellow-cedar cutting-donor. **b** Yellow-cedar 2-year old container cutting-orchard. A 450 m^2 green-house will hold 10000 cutting-donors in 11-1 pots, and will potentially produce 500000 3- to 5-cm cuttings annually for up to 5 years without repotting.

spacing, and proper watering and fertilization regimes (Fig. 5) (Russell et al. 1990).

There are, however, advantages to using stecklings originating from stage-II donors with respect to deer browsing, cold tolerance, and genetic improvement. Deer preferably browse seedlings over stecklings, and stecklings from stage-I donors are more prone to browsing than are stage-II stecklings. Similarly, 1-year-old seedlings are more susceptible to cold damage than are stage-I stecklings, which are more prone to cold damage than are stage-II stecklings (G. Dunsworth unpubl. data, MacMillan Bloedel, Nanaimo, B.C).

Finally, using stage-II cuttings will be necessary in order to realize the genetic and management gains possible from clonal forestry if hedge-origin stecklings are the propagule type (See Sect. 6, below). The increased nursery costs from growing stage-II stecklings should be easily offset by the advantages of clonal forestry and some of the favorable tradeoffs (browse, frost, etc) associated with the more mature stecklings. Additionally, favorable results to date with serial propagation indicate that cutting-donors can be maintained in stage-I maturation chronologically longer than with hedging.

5 Seedling:Steckling Comparisons

Results from comparisons among seedlings and several maturation levels of stecklings in the nursery, during field establishment, and in field trials, have been encouraging.

5.1 Nursery Comparisons

One-year-old stecklings originating from hedged cutting-donors as old as 13 years from seed behave similarly to 1-year-old seedlings with respect to morphological traits including height growth, shoot form, root-collar diameter, root dry-weight, and root-growth capacity, and in most of these traits the stecklings were better than the seedlings (Fig. 6). As well, stecklings showed

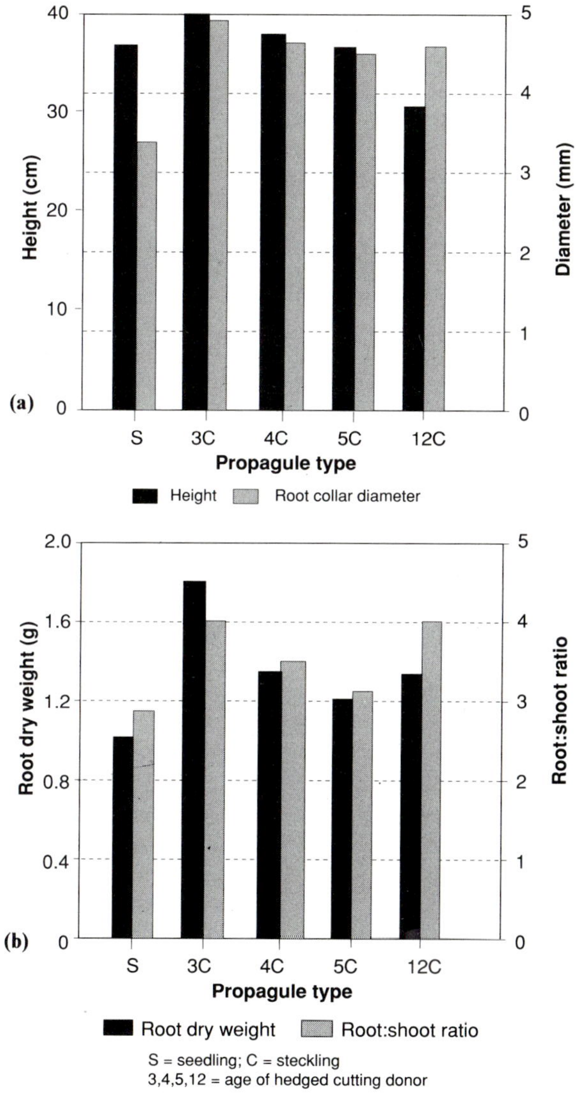

Fig. 6a,b. Effect of hedged cutting-donor age (seedling-origin) on the performance of 1-year-old stecklings, and comparison to 1-year-old seedlings, **a** Height and root collar diameter, **b** Root dry weight and root:shoot dry weight ratio. All seedlings and stecklings were from the same seedlot, and were grown in the same environment and container type. (Russell et al. 1990).

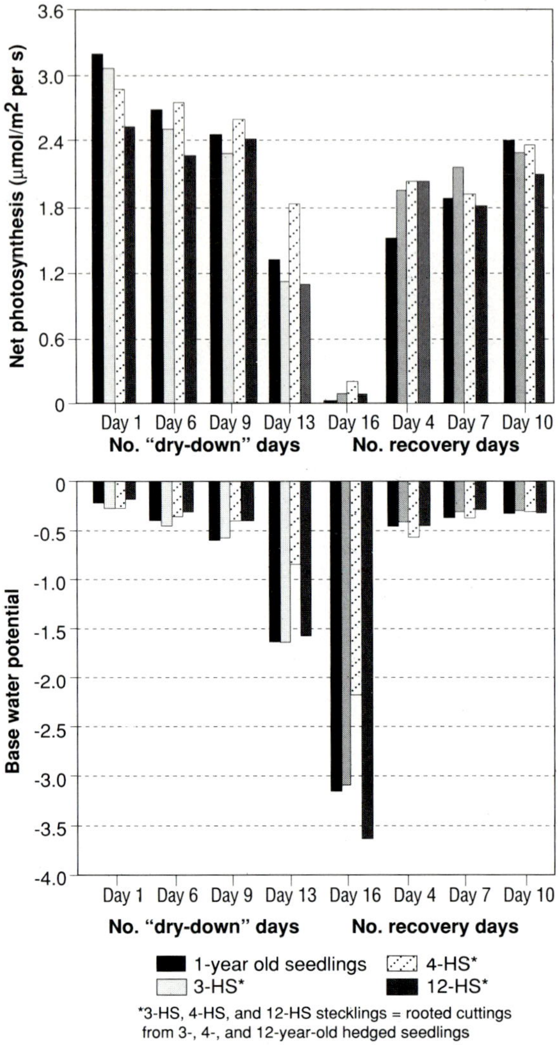

Fig. 7. Physiological response to drought of 1-year-old seedlings and stecklings originating from hedged cutting-donors of three different chronological ages. (Russell et al. 1990)

physiological responses including base water potential, net photosynthesis, stomatal conductance, transpiration, and root growth capacity similar to those of seedlings during a simulated drought period and recovery phase in a growth room (Fig. 7) (Grossnickle and Russell, unpubl. data).

5.2 Field Comparisons

Two field trials comparing yellow-cedar stecklings and seedlings were planted on typical reforestation sites for the species (Karlsson 1981). The first trial, planted in the fall of 1976, compared seedlings with stecklings from 1- and 3-year-old nonhedged donors, and from 7-year-old hedged donors. The second trial, planted in the fall of 1978, compared seedlings with stecklings from 2-year-old cutting donors.

These two trials were measured in the fall of 1988 for survival, height, diameter, and form (Karlsson and Russell 1990). There were no significant differences between trees of seedling origin and any of the steckling origins, for any of the traits measured in either trial. Furthermore, there was no evidence of sustained plagiotropic growth among any of the trees of steckling origin.

In a farm-field trial comparing seedlings and stecklings from 25 cm hedged and unhedged 11-year cutting-donors, there were no significant differences in fourth-year height (Russell and Ferguson unpubl. data). However, stecklings from unhedged donors exhibited a higher degree of sustained plagiotropism. It is perhaps surprising that stecklings from unhedged donors performed as well as stecklings from hedged donors and seedlings with respect to height growth. However, rooted cuttings were culled at the nursery stage based on height, root-collar diameter, and root mass, (there was an increase in cull rate with an increase in donor hedging height) and only better-quality stock was planted. Thus, although hedging of cutting-donors increases steckling stock quality, and thus decreases cull rate, cuttings from even the upper crowns of 11-year-old trees that do not get culled at the nursery stage grow as well as stecklings originating from hedged donors.

Results from these comparisons indicate that hedging of cutting-donors up to at least 13 years delays maturation such that the resultant stecklings compare favorably to seedlings with respect to nursery production and stock quality, field establishment, and early field performance. No field trials have yet been established for serial-propagated stecklings.

6 Genetic Improvement

The yellow-cedar genetic improvement program involves two stages: (1) short-term genetic gain through clonal testing and selection and; (2) long-term genetic gain involving both clonal selection and recurrent selection for additive genetic variation.

6.1 Short-Term Genetic Gain Through Clonal Testing

Clonal forestry for yellow-cedar began in 1986 with the initiation of two separate but complimentary strategies developed by two different agencies in British Columbia. The first strategy involves farm-field candidacy testing (Libby 1987) and the establishment of clonal hedges, and the second strategy involves the roguing of an existing seedling cutting-orchard through clonal testing on reforestation sites (J. Barker, Western Forest Products, pers comm 1990).

6.1.1 Farm-Field Candidacy Tests and Clonal Cutting-Orchard

This strategy involves farm-field clonal testing (candidacy testing; Libby 1987) of open-pollinated family individuals and the concurrent establishment of clonal cutting-donors (Russell 1988). One-year-old seedlings are cloned, with six ramets per clone established in a farm-field trial using noncontiguous plots with three blocks and two interlocking replications. The sites are well prepared, trees planted at close spacings, watered as needed, and competing brush controlled. A container clonal cutting-orchard is established concurrent with the clonal tests with four to six ramets per clone. There are approximately 1000 clones in tests.

The cutting-orchard will be initially rogued based on growth, wood density, and form at age 5, mass selecting the top 10% of the clones with a restriction of three clones per open-pollinated family. The selected clones will be vegetatively multiplied for production, using a combination of serial propagation and hedging. A further roguing at a later age to the top 5% may be done, depending on the outcome of future maturation-related field trials. Although this strategy will result in early genetic gains, potential problems with clone-by-site interactions and the hope that favorable substantial juvenile:mature correlations exist may substantially decrease gains.

6.1.2 Roguing Existing Seedling Cutting-Orchards

This strategy involves the cloning of individuals from an established seedling cutting-orchard (4- to 6-years-old) and testing on a number of sites (J. Barker, Western Forest Products, pers. commun 1990) with the objective of roguing the existing orchard. Again, this strategy skips the initial screening stage (Libby 1987) and proceeds with candidacy testing.

Clones are being deployed to four reforestation sites with nine ramets per clone per site planted in noncontiguous plots. An initial selection will be done on rootability, with poor-rooting clones being rogued out of the existing cutting-orchard. Clones will then be selected based on early form and wood quality at 6 to 8 years and the existing cutting-orchard then rogued. Currently, there are 300 clones in tests, with up to 2000 planned for the future.

6.2 Long-Term Genetic Gain

The long-term genetic improvement strategy for yellow-cedar involves (1) recurrent selection for additive genetic variation through repeated selection and breeding, and (2) deployment of genetic gain through propagation of selected clones. A population hierarchy will be developed with the production population being a subset of the breeding population, which is a subset of the gene pool population. The purpose of the gene pool population is long-term preservation of selected and unselected genotypes. The purpose of the breeding population is for recombination of genes, and for the selection of additive genetic variation. The purpose of the production population is for the deployment of genetically improved clones for reforestation. In order to implement an efficient program and to maximize genetic gain per unit time, the above objectives for the breeding and production populations will be combined into one test by clonally replicating seedlings from controlled matings. This will allow for a dual purpose test: selection of an advanced-generation breeding population based on additive gene effects, and selection of clones for the production population. This strategy has the potential to maximize total genetic gain (Shaw and Hood 1985; Shelbourne 1990), as well as genetic gain per unit time (Matheson and Lindgren 1985).

Preliminary results from a genetic architecture study (Russell, unpubl. data) indicates the feasibility of using one breeding zone, which encompasses 90% of future areas to be reforested with yellow-cedar. The breeding population for this zone currently consists of 192 phenotypic selections from wild stands established in a breeding orchard. A mating design involving eight-clone disconnected circular partial diallels (Lowe and van Bujitenan 1986) will be used to produce full-sib progeny. Each diallel will result in 16 full-sib families for a total of 384 full-sib families. Four series of breeding, testing, and selection will be done, with six diallels per series.

Progeny testing for ranking parents based on GCA and clonal testing for selecting superior clones based on genotypic value (clonal mean) will be combined as mentioned above. A proposed design involves the cloning of 20 1-year-old seedlings per full-sib family and deploying two ramets per clone on each of four sites. This would result in 3840 trees, planted on 1.5 ha per site with 2 m^2 spacing. Clones will be deployed using sets within replications with full-sib families balanced across all replications within a site (Foster and Shaw 1987). This strategy will result in the testing of 1920 clones per series for a total of 7680 clones over four series.

The above design is a compromise between numbers of families, clones, and ramets tested with respect to gain in both the production and breeding population. The optimal distribution of effort, assuming a fixed-resources scenario, depends upon many factors including genetic and phenotypic variances, genotype × environment interactions, selection intensity, and selection strategy. Based on previous and ongoing theoretical studies, the above strategy seems reasonable with respect to maximizing gain in both populations (Shaw and Hood 1985; Russell and Libby 1986; Russell and Loo-Dinkins in preparation).

Selections for the production population will be deployed operationally using rooted cuttings. The top 25 clones per series will be selected at age 12 based on growth and wood quality. A restriction of three clones per full-sib family and six clones per half-sib family will be placed on the selections. Combined index selection may be used for clonal selection; however, since coancestry will be controlled, simple mass clonal selection may yield similar results. This has yet to be fully explored and will depend upon levels of coancestry control, proportional amounts of additive and nonadditive genetic variance and selection intensities.

At the time of progeny planting, a cutting-orchard will be established with one ramet per clone transplanted into 4-l pots and pruned annually for the maintenance of juvenility. A 50% roguing of the progeny-tested clones will be done in the cutting-orchard based on 4-year data. Each of the remaining selected clones will be repropagated to further maintain juvenility and to multiply each clone to two ramets. A similar roguing will be done at age 8 with another 50% reduction in clones in the cutting-orchard and a repropagating of selected clones and a doubling of ramets to four per clone. At age 12, the best 25 clones per series will be selected, repropagated, and vegetatively multiplied, and within 2 years of selection the cutting-orchard will be in full production (10 000 stecklings per clone × 25 clones/series × 4 series = 1 million stecklings). Successive roguings at 4-year intervals serves two purposes: (1) maintenance of juvenility through a combination of serial propagation and pruning, and (2) producing a rogued orchard at production scale levels soon after final selection age.

Selections for an advanced-generation breeding population will be based on the same traits, and at the same age as the production selections. Again, given coancestrial control, optimal gains with clonal replicates may not be achieved with index selection but with three-stage tandem selection (Shaw and Hood 1985). Eight clones per diallel for a total of 192 clones will be selected for the second-generation breeding population.

7 Summary

Historically, a steckling production system was developed for yellow-cedar in British Columbia in order to fulfill a demand for planting stock. The results to date are an efficient, large-scale steckling production system and a good understanding of cutting-donor maturation and its effects on steckling quality. There is over 15 years' experience in growing yellow-cedar stecklings. Currently, over half a million stecklings are planted annually on reforestation sites with encouraging results. The oldest yellow-cedar steckling research plot is now 13 years old and, to date, the survival, growth, and form of stecklings are similar to seedlings. Thus, we are gaining knowledge and confidence in the use of stecklings for reforestation.

An efficient and economical nursery-production system for producing yellow-cedar stecklings is in place, and has successfully been transferred to the private sector. The knowledge and techniques are now available for controlling and manipulating cutting maturity and physiology. Research has increased our understanding of physiological responses associated with maturation. This will result in a better understanding of the processes necessary to produce quality stecklings.

Over 1000 clones are currently being candidacy-tested with plans for a total of over 3000 candidacy clones in the short-term gain program. These will be rogued to a maximum of 200 selected clones, for production within 5 years. Initial results are encouraging with respect to levels of clonal variability in the farm-field trials.

Finally, a breeding strategy involving recurrent selection for additive genetic variance, and the deployment of selected clones is in place. Selected clones will be maintained in a potted cutting-orchard and vegetatively multiplied through serial propagation. However, future propagation techniques should enhance the potential of yellow-cedar clonal forestry. Active research programs involving organogenesis and somatic embryogenesis have provided encouraging results and may either supplement or replace stecklings in the future.

References

Andersen HE (1959) Silvical characteristics of Alaska-cedar (*Chamaecyparis nootkatensis*). **USDA** Alaska For Res Stn Pap 11: 10 pp

Foster GS, Shaw DV (1987) A tree improvement program to develop clones of loblolly pine for reforestation. In: Proc 19th **SFTIC** Coll Stn, Texas, pp 17–21

Grossnickle SC, Russell JH (1989) Rooting of yellow cypress cuttings, ptl: Influence of donor plant maturation. BC FRDA. Memo 083:2 pp

Karlsson I (1974) Rooted cuttings of yellow cedar. BCFS Int Res Note 66:5 pp

Karlsson I (1981) Propagation of Alaska yellow cedar (*Chamaecyparis nootkatensis* (D. Don) Spach.) by rooted cuttings for production plantings. Proc Int Plant Propagators Soc 31:112–116

Karlsson I, Russell JH (1990) Comparison of yellow cypress trees of seedling and rooted cutting origins after 9 and 11 years in the field. Can J For Res 20:37–42

Libby WJ (1987) Testing for clonal forestry. Ann Forest 13(1–2):69–75

Lowe WJ, van Buijtenan JP (1986) The development of a sublining system in an operational tree improvement program. In: Proc IUFRO Meet Working Party Breeding, progeny testing, and seed orchards, Williamsburg, VA, pp 98–106

Matheson AC, Lindgren D (1985) Gains from the clonal and the clonal seed-orchard options compared for tree breeding programs. Theor Appl Genet 71:242–249

Russell JH (1988) The use of rooted cuttings in British Columbia's tree improvement and silvicultural program. Comb Proc Int Plant Propagators Soc 38:151–155

Russell JH, Libby WJ (1986) Clonal testing efficiency: the tradeoffs between clones tested and ramets per clone. Can J For Res 16:925–930

Russell JH, Loo-Dinkins J (accepted by Silvae Genetica) Distribution of testing effort among computer-simulated clonal populations.

Russell JH, Grossnickle SC, Ferguson C, Carson DW (1990) Yellow-cedar stecklings: nursery production and field performance. BC FRDA Rep 184:20 pp

Shaw DV, Hood JV (1985) Maximizing genetic gain per effort by using clonal replicated in genetic tests. Theor Appl Genet 71:392–399

Shelbourne CJA (1990) Incorporating clonal forestry in the breeding strategy. In: Proc Joint Meet WFGA and IUFRO Working Parties S2. 02–05, 06, 12, and 14, Olympia, WA

12 Organization and Implementation of a Clonal Forestry Program

S. Strobl and R.W. Evers[1]

1 Profile of an Operational Clonal Forestry Program

1.1 Rationale for Program Beginnings

A successful large-scale operational clonal forestry program has evolved in eastern Ontario, Canada since 1975. The beginning of this program can be attributed to three main factors: (1) the demonstrated growth of improved hybrid poplar clones in field tests, (2) the developing wood supply shortage of a local pulp and paper mill, and (3) the availability of a large surplus of idle agricultural land suitable for reforestation with hybrid poplar.

The Ontario Ministry of Natural Resources (OMNR) had been supporting a poplar breeding and clonal selection program led by Dr. Louis Zsuffa, then a forest geneticist with the OMNR's research group. Since the early 1960s, imported clones from traditional European poplar culture, as well as selections from the progeny of artificial crosses between native eastern cottonwood (*Populus deltoides*) and imported pollen sources, were screened in field tests throughout Ontario (Zsuffa 1979). Field tests in eastern Ontario demonstrated a great potential for growing and harvesting hybrid poplar for wood fiber production.

At the same time, a large fine-paper mill in the area was experiencing difficulties obtaining an adequate hardwood feedstock supply. Domtar Inc., of Cornwall, Ontario, was importing 65% of its wood several hundred kilometers from northern New York State and the province of Quebec. Subsequently, the mill managers were eager to start an operational hybrid poplar clonal reforestation program with the long-term goal of eventually securing a part of their wood supply needs close to the mill.

Finally, due to changes in the farming sector, a large surplus (0.6 million ha) of idle marginal agricultural land had been abandoned in the eastern region (Wayman 1978). Of this land base, 0.4 million ha was identified as suitable for reforestation with hybrid poplar.

It was these initial factors which led to the establishment of an operational clonal forestry program in eastern Ontario. The present-day success of this program, however, would not be possible if it were not for the close link between technology development and transfer, and operations. Technology development

[1]Fast Growing Forests, Ontario Ministry of Natural Resources, Box 605, Brockville, Ontario K6V 5Y8, Canada

Clonal Forestry II, Conservation and Application
Ed. by M.R. Ahuja and W.J. Libby

activities support innovation in poplar breeding and clonal selection, cultural practices (i.e., site preparation, tending, and fertilization), and the development of production systems (e.g., crop density and rotation). A technology transfer program is responsible for ensuring that these innovations are applied to the operational program.

1.2 Synopsis of the Current Operational Program

Each year Domtar Inc. plants approximately 300 000 cuttings of some 25 clones of hybrid poplar currently considered as production clones. Since 1975, approximately 2260 ha of hybrid poplar plantations have been established within a 25-km radius of the mill in Cornwall, Ontario (see Fig. 1). The company's goal is to eventually supply 20% of their mill furnish from hybrid poplar plantations. Currently, less than 1% of the mill's annual need for 225 000 oven-dry tons of hardwood feedstock is met by existing plantations.

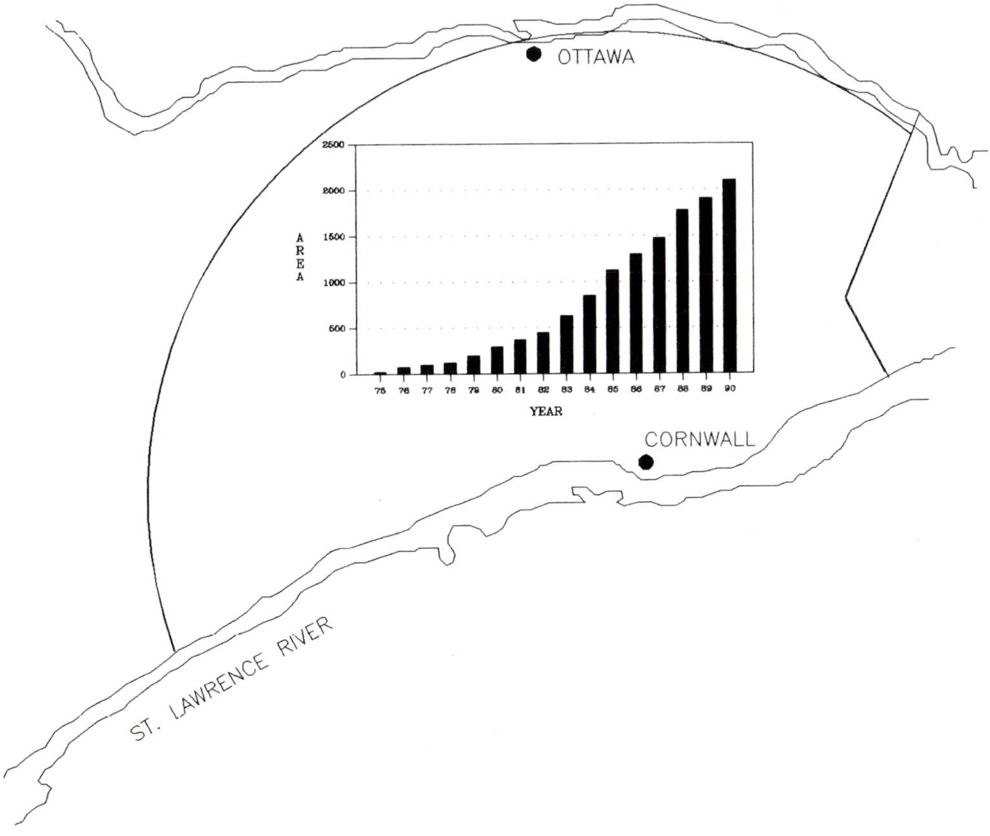

Fig. 1. Location of hybrid poplar program in eastern Ontario and cumulative area planted to hybrid poplar during the period 1975–1990

Before planting begins, the soil of prospective planting sites is sampled at regular intervals along parallel transects. Soil types are differentiated on the basis of soil texture, depth, drainage and stoniness. Fairly homogenous compartments are delineated on the property map based on soil type. The soil survey serves as the basis for later site preparation, tending, and clonal allocation decisions (Evers et al. 1983).

An application of the herbicide glyphosate, deep plowing and discing remove existing vegetation to prepare the typically old field sites for planting. The site is then cultivated before application of the pre-emergent herbicide, simazine. In the spring, 25 cm cuttings are planted by hand at a 3.6×2.4 m spacing (or 1100 stems/ha). Plantations are tended for about 3 years until they are determined to be free to grow, after which time they are maintenance-free until harvest at age 12.

During the winters of 1988 and 1989, the first operational harvests of hybrid poplar took place in eastern Ontario. Sixty-three hectares were harvested with an average productivity of 2.54 oven-dry metric tons (ODT)/ha/yr or 5.8 m^3/ha/yr. The variability in productivity was large, however, ranging from 3.49 m^3/ha/yr to 11.19 m^3/ha/yr (or 1.66 ODT/ha/yr to 4.55 ODT/ha/yr). This variability can be attributed to two factors: (1) poor site preparation and tending in the early years of the program, resulting in less than adequate stocking, and (2) too few clones suited to climatic conditions in eastern Ontario. Productivity is expected to increase to a production goal of 4.0 ODT/ha/yr as we begin to harvest plantations established with better site preparation and tending technologies and more intensively tested clones.

1.3 Key Partners

Over the past 15 years, many organizations and people have been involved with the development of the hybrid poplar operational clonal forestry program in eastern Ontario. Research support by Dr. Zsuffa has already been mentioned, and other Canadian, American, and international scientists have also contributed to the many research needs of the program.

The technology development and transfer role has been filled by the Fast Growing Forests (FGF) group of the OMNR. The group has formed partnerships with corporate, public, and governmental agencies, to develop and transfer the intensive forest management technologies associated with the hybrid poplar clonal forestry program. There are now several private and corporate growers of hybrid poplar besides Domtar Inc. in Ontario. Nursery staff have also played an instrumental role in developing large-scale technologies to grow and produce several hundred thousand hybrid poplar cuttings annually. Technology development work has been jointly funded by the Federal and Provincial governments.

Finally, field staff with the OMNR and Domtar Inc. provide a continuing effort to maintain consistent growth and survival in plantations as the scale of the program increases. Field staff are involved in either of two operational programs which exist for the establishment of plantations.

In the first program, the company buys the land and OMNR staff manage these lands under agreement, i.e., they are responsible for annual work plans which include site preparation, planting, tending, and harvesting operations. The OMNR is partially reimbursed through stumpage charges when the wood is commercially harvested.

Under the second program, the company leases land from local farmers through a cost-shared program with the OMNR, very similar to Forest Management Agreements in northern Ontario. The company has the first right of refusal on the wood at rotation and will pay the landowner the current rate for stumpage. The company supplies staff to manage these leased lands.

2 Evolution of the Current Program

The current hybrid poplar program as described has evolved since 1975. Two areas of the program which have seen the greatest change over this 15-year period are (1) clonal testing and deployment strategies and (2) plantation establishment technologies, including soil surveying, site preparation, and tending.

2.1 Clonal Management

2.1.1 Evolution of Production Clones

The most substantial advances have been achieved through testing and selection of hybrid poplar clones. Table 1 documents the changes to the production clone list from the period 1976 to 1988. In 1976, for example, only nine clones were planted, and six of these later proved to be ill-adapted to eastern Ontario's harsh winter.

The first candidates for clonal testing included promising clones selected from clonal screening tests in eastern Ontario established in the early 1970s by Dr. Zsuffa. During these years, primarily clones of Aigeiros (i.e., *P. deltoides*

Table 1. Hybrid poplar production clones in eastern Ontario during the period 1976-1988

	Clone	1976	1980	1984	1988
AK30	*P. alba* A69 × *glandulosa* GL1			*	
AK41	*P. alba* A69 × *glandulosa* GL1			*	
DN1	*P. euramericana* cv. Allestein	*	*	*	*
DN2	*P. euramericana* cv. Baden		*		*
DN3	*P. euramericana* cv. Dolomiten		*		
DN5	*P. euramericana* cv. Gelrica		*		*
DN9	*P. euramericana* cv. Lons	*			
DN10	*P. euramericana* cv. Robusta		*		*
DN16	*P. euramericana* cv. Regenerata Batarde d'Hauterive	*	*	*	*

Table 1 continued

	Clone	1976	1980	1984	1988
DN17	*P. euramericana* cv. Robusta	*		*	*
DN19	*P. euramericana* cv. Blanc di Poitou		*	*	
DN21	*P. euramericana* cv. I-78B (LW 42)	*			
DN22	*P. euramericana* cv. I-262		*		
DN31	*P. euramericana* cv. Negrito de Granda	*			
DN33	*P. euramericana* cv. Virginie Nancy		*	*	*
DN38	*P. deltoides* Virginiana × Volga Poplar (NE-238)	*	*	*	
DN55	*P. deltoides* D39 × *nigra* N3, Pop 558-1		*		
DN70	*P. deltoides* D35 × *nigra* N3, Pop 537-1		*		*
DN74	*P. deltoides* D35 × *nigra* N3, Pop 537-5		*	*	*
DN93	*P. euramericana* cv. Triplo				*
DN106	*P. deltoides* D36 × *nigra* N1 + 2 mixed Pop 683-27		*	*	
DN114	*P. deltoides* D58 × *nigra* *N*1 + 2 mixed Pop 694-5				*
DN125	*P. deltoides* D58 × *nigra* N1 + 2 mixed Pop 694-28			*	
DN128	*P. deltoides* D56 × *nigra* N1 + 2 mixed Pop 696-3		*	*	*
DN131	*P. deltoides* D56 × *nigra* N1 + 2 mixed Pop 696-7			*	
DN132	*P. deltoides* D56 × *nigra* N1 + 2 used Pop 696-8				*
DN134	*P. deltoides* D56 × *nigra* N1 + 2 mixed Pop 696-10		*		
DN152	*P. deltoides* D56 × *nigra* N1 + 2 mixed Pop 696-29				*
DN157	*P. deltoides* D36 × *nigra* N1 + 2 mixed Pop 683-45				*
DN160	*P. deltoides* D36 × *nigra* N1 + 2 mixed Pop 683-48				*
DN174	*P. euramericana* cv. Flevo (Koster 923)				*
DN177	*P. euramericana* cv. Spijk (Koster 2195)				*
DN178	*P. euramericana* cv. Spijk (Koster 2205)			*	
DN181	*P. euramericana* cv. Drapal Wulst			*	*
DTAC1	*P. angulata* × *berolinensis* (NE 32)	*	*		
DTAC2	*P. angulata* × *berolinensis* (NE 32)		*		
DTAC32	*P. deltoides* × *balsamifera*			*	
GA88	*P. grandidentata* G321 × *alba* A321			*	
IH45/51	*P. euramericana*	*	*		
Jac4	*P. balsamifera* × *deltoides*		*	*	
Jac31	*P. deltoides* cv. Roller # F1		*		
Jac34	*P. maximowiczii* × *trichocarpa*		*	*	
JacN5	*P. jackii* × *nigra* N1 + 2		*	*	
NM1	*P. nigra* × *maximowiczii*			*	
NM2	*P. nigra* × *maximowiczii* cv. Max-1			*	
NM3	*P. nigra* × *maximowiczii* cv. Max-2			*	
NM4	*P. nigra* × *maximowiczii* cv. Max-3			*	
NM5	*P. nigra* × *maximowiczii* cv. Max-4			*	
NM6	*P. nigra* × *maximowiczii* cv. Max-5			*	*
TOTAL		9	23	26	20

×*nigra* or *P.*×*euramericana* hybrids) origin were screened for adaption to eastern Ontario soil and climatic conditions.

However, in 1981, an international committee of poplar breeders who reviewed the program recommended that clones with balsam (Section Tacamahaca) parentage, both exotic and native (e.g., *P*×*jackii*) species hybrids, should also be screened. European poplar breeders had obtained substantial growth increases with hybrid poplar clones of Tacamahaca parentage (notably *P. trichocarpa*×*deltoides* hybrids), especially on heavy soils, as compared to traditional *P.*×*euramericana* hybrids (Tessier du Cros 1984).

The European observation of faster growth with Tacamahaca hybrids was quickly verified in eastern Ontario with six *P. nigra*×*maximowiczii* (NM) hybrids originally selected in Japan. During the period 1981 to 1986, approximately 203 ha of plantations in the Domtar woodshed were established with these six clones.

By 1987, however, a high incidence of Septoria canker was observed in plantations of these six exotic NM clones. Septoria canker is caused by *Septoria musiva*, an endemic pathogen that causes both leaf spots and cankers on hybrid poplar clones. Five of the clones were found to be too susceptible for further plantation establishment and were subsequently eliminated from the production clone list (see Fig. 2). A survey of canker incidence in these plantations found that 79% of the area was affected to some degree with Septoria damage (Strobl and Fraser 1989). Several other clones have subsequently been found to be too susceptible to Septoria canker and are no longer planted, including *P.*×*jackii* clones and a few *P. deltoides*×*nigra* hybrids.

This recent experience precipitated a change in the clonal testing strategy for new clones. It was apparent that local screening for resistance to Septoria canker was the most promising long-term control possibility and early screening tests

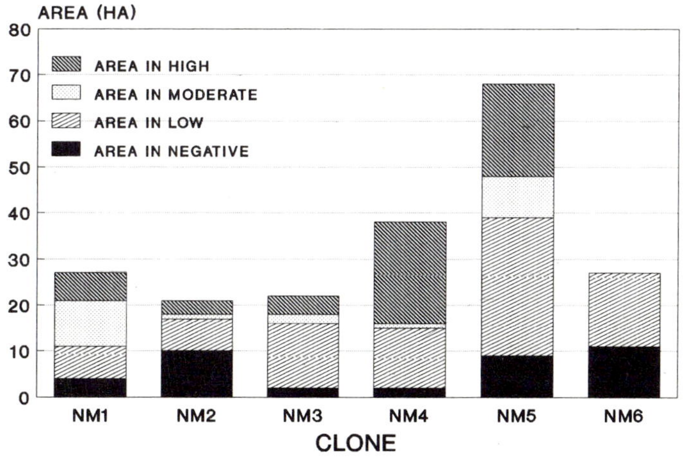

Fig. 2. Area affected by *Septoria* canker in each severity class by clone. (After Strobl and Fraser 1989)

(i.e., prior to field testing) needed to be incorporated in the clonal testing strategy.

2.1.2 Evolution of the Clonal Testing Strategy

Previously, clonal field tests were an integral component of the hybrid poplar clonal forestry program in eastern Ontario. During the period 1976 to 1986, 120 clone-site trials (CSTs) were established on 197 ha. Approximately 486 clones were tested in CSTs during this period, resulting in the identification of production clones as shown in Table 1. Results of this decade of clone-site testing are documented in Strobl (1988).

CSTs were established to determine the survival and performance of different clones on the various production site types which exist in the Domtar woodshed. The allocation of clones to large-scale production plantations involves the matching of tested clones to new planting sites of known soil types. Careful matching of clone to site ensures that each clone is planted within the range of sites upon which it grows best (Evers et al. 1983).

It was thought that the testing process could be improved by including an early screening stage at the nursery, thereby eliminating expensive field testing of clones with poor rooting ability and susceptibility to diseases, especially *Septoria* canker. Subsequently, a three-stage testing program, which closely parallels the testing program used at the Belgian Poplar Research Center (Steenackers and Strobl 1988), was introduced to screen new clones.

The first stage involves a simple nursery screening test which need not be replicated as nursery conditions can be considered uniform. The objective of this test is to identify those clones which are (1) resistant to *Septoria* canker, *Melampsora* leaf rust, and frost, and have (2) good rooting ability and (3) fast growth. A reliable, early artificial inoculation test is being developed to screen clones for resistance to *Septoria* canker. The practice of keeping plants in the nursery for an extended period, during which better knowledge of the clones can be obtained at lower cost, permits the preliminary selection of more valuable material (Muhle Larsen 1970). Consequently, further efforts are concentrated on the most promising material.

In the second stage of testing, the most promising clones from the nursery screening test are put into field tests on a range of different sites and over a period of establishment years. Spearman rank correlations have shown close to perfect positive rank correlation between 5th year and final year rankings for both height and diameter, indicating that it is not necessary to wait for 10-year data to identify top-ranking clones (Strobl 1988). Clones are tested as a group on several different sites. In the past, CSTs were established with different numbers and groups of clones, making comparisons of clonal performance over a range of sites and calculation of genotype x environment interactions difficult.

The objective of the CSTs is to identify both plastic and site-specific clones. For site-specific clones, it is necessary to establish on which sites a particular clone performs best. Increasingly, it is becoming apparent that planting exclusively plastic clones may be advantageous on the very spatially heterogeneous sites which exist in eastern Ontario, moreover, operational considerations favour the deployment of clones adapted to a wide variety of site types. However, it is well recognized that a deployment strategy which favors plastic clones may come at some cost to optimal productivity because clones adapted to specific site types may achieve greater productivity on these sites than plastic clones (Strobl 1988).

The third stage of testing involves the establishment of larger monoclonal blocks for the estimation of yields as suggested by Libby (1987). The most promising clones from CSTs are put into production yield trials of 192-tree plots (i.e., 16 × 12 configuration) with only the inner 12 × 8 trees being measured. This test will provide better unit-area productivity estimates than those which can be obtained from four-tree plot CSTs. Moreover, this test provides a final opportunity to assess candidate clones before their possible release to cultivation.

This three-stage testing strategy is particularly important now that we are breeding our own clones rather than relying exclusively on importation of clones from poplar institutes throughout the world. Possibilities for further importation of poplar clones from elsewhere have been virtually exhausted.

For example, 165 clones were recently received through an International Energy Agency (IEA) agreement with the Poplar Research Center at Grammont, Belgium. Following extensive nursery screening over a 3-year period, only 34 clones could be selected for further field testing. It is not surprising that most of these clones were not suited to eastern Ontario's harsher climate (see Table 2) and/or were too susceptible to our species of *Melampsora* rust. These clones were originally selected for a milder continental climate in Europe with different disease problems, thereby limiting the possibilities for selection in other conditions (Steenackers et al. 1990).

To support the scale of planting of hybrid poplar by Domtar Inc. as well as other growers in Ontario, new clones must continually be developed. Since 1987, poplar breeding efforts by the FGF group have concentrated on the production of first generation hybrid populations from which to select individuals for systematic testing in clonal screening trials. Hybrid crosses are being made between native *P. deltoides* female plus tree selections from eastern Ontario and imported pollen sources, primarily *P. nigra*, *P. maximowiczii*, and *P. trichocarpa*.

The immediate goal of this breeding program is to develop superior hybrid poplar clones with eastern Ontario *P. deltoides* parentage – this species being one of the most important parent species of hybrid poplar clones, both in Europe and in North America. More than 500 plus tree selections of *P. deltoides* have been made in the eastern region and many of these have been preserved in breeding arboreta. Experience with clones screened from other programs have

Table 2. Climate of Eastern Ontario

	Latitude longitude	Elevation (m)	Average annual temp. (°C)	Average warmest temp. (°C) (July)	Average coolest temp. (°C) (Jan.)	Average annual high/low temp. (°C)	Annual precipitation (mm)	Growing season no. of days above 5.6°C	No. of days of precipiation per year
Climate of Ottawa	45°25′N 75°42′W	87.0	5.3	20.3	− 11.1	34.0 − 31.0	852	190	60

identified our most promising hybrids as *P. deltoides* × *maximowiczii* and *P.* × *euramericana* crosses.

The best growth – upwards to 5.0 ODT/ha/yr in some CSTs – has been obtained with hybrids in which *P. maximowiczii* has been one of the parents. This balsam poplar, a native to eastern China, Korea, and Japan, tolerates a wide range of sites and does especially well on the typically marginal sites in eastern Ontario. However, this species is extremely susceptible to *Septoria* canker and has a low wood density. Major gains can be anticipated from combining it with *P. deltoides*, which is both resistant to the canker and has a higher wood density.

The other hybrid cross that is of interest is the traditional *P.* × *euramericana*. These hybrids are typically more nutrient-demanding, but have excellent wood quality and higher canker resistance. They will have broad applicability to the better-quality agricultural land which is available to Domtar Inc. through the land-lease program. The majority of our current production clones belong to *P.* × *euramericana*.

2.1.3 Evolution of the Clonal Deployment Strategy

The rationale for clonal deployment to sites has also changed over time. In the early years, each planting site was subdivided into fairly homogenous blocks based on soil survey information. Data on clonal performance from CSTs established on similar soils were then used to determine the most suitable clones for each block. The maximum size allocated per clone was 2 ha per property, and although a number of the blocks may have been of the same soil type, a different clone was allocated to each block. It was felt that this deployment strategy resulted in a high level of genetic diversity, and any pest problems which developed would be localized to a small stand on a given property (Barkley et al. 1983b).

Operationally, however, the practice of restricting clonal blocks to a minimum size of 2 ha was costly. Field staff required a more flexible clonal deployment strategy. For example, it was not cost-efficient to plant two clones of 3 ha each on a fairly uniform 6 ha field. Moreover, a localized pest problem in a small stand on a given property was extremely expensive to correct due to economies of scale.

With the new three-stage testing process the field forester's concerns for operational efficiency (given that the planting site is uniform and/or the allocated clone is plastic) and the scientist's concerns for genetic diversity can both be satisfied. By looking at new clones over a period of several years and in varying conditions (i.e., nursery row plot, CST test, and monoclonal stand), greater knowledge regarding a candidate clone's adaption to our climate and soils, and its resistance to our prevalent diseases can be obtained. Protection from unforseen events is assured by deploying a large number of diverse clones annually and by adding new clones over time. In our program, genetic diversity is now controlled at the nursery – a maximum of 20000 cuttings of each of our current 25 production clones are produced annually. Also, new clones are added to the production list annually, while older clones are removed.

We are comfortable with our chosen clonal deployment strategy. Our strategy's merits were recently demonstrated as we faced an escalating pest problem with *Septoria* canker, which required the premature harvesting of 2.5% (38.7 ha) of the total area planted to hybrid poplar in eastern Ontario. This measure was necessary to reduce inoculum levels and slow infection to nearby plantations of other susceptible clones (Strobl and Fraser 1989). We feel that we can accept such a level of failure from an unforeseen event. By having a large number of production clones with wide genetic diversity, we were able to meet that year's stock requirements despite having to remove all susceptible clones from our inventories.

2.2 Plantation Establishment Technologies

Concurrent to efficient management of genetic diversity, a successful operational clonal forestry program ensures that the inherent potential of production clones is not compromised by deployment to inadequate sites. Growth and survival in early plantations was extremely variable, and closer examination determined that this variation was due to both differences in site characteristics and cultural inputs. The importance of proper site selection to identify suitable sites for hybrid poplar and to allocate site-specific clones was quickly realized. Moreover, proper site preparation and tending were found to be instrumental to assuring maximum growth of selected clonal material.

2.2.1 Soil Survey Methodology

Soil survey standards for hybrid poplar were developed in the early years of the program and have been extensively documented (Barkley et al. 1983a; Evers et al. 1983, 1984). Soil characteristics of potential plantation sites are sampled at a scale of 1:2000 and soil types are differentiated on the basis of soil texture, depth, drainage, stoniness, etc. A final map delineating all distinct soil types is produced at a scale of 1:4000. This map forms the basis for subsequent management decisions, including site preparation modifications, clonal allocation, tending prescriptions, and fertilization requirements. After 1983, the soil survey was contracted out to commercial consulting firms in an effort to make this procedure an integral component of plantation establishment. Since then the soil sampling intensity has changed to 1:10000 as it became apparent that it was only necessary to capture major soil trends on planting sites, primarily changes in drainage which have been determined to have the greatest influence on poplar growth (Rogers et al. 1984).

2.2.2 Cultural Practices

Thorough site preparation to eliminate all competition is important for the successful establishment of hybrid poplar plantations (Raitanen et al. 1983). In the early years, site preparation was primarily achieved by mechanical discing. However, discing only exacerbated problems with quack grass and other deep-rooted perennials common to eastern Ontario's old field sites. Examination of the existing weed control literature suggested two possibilities – glyphosate, a relatively new chemical at the time, and EPTC, a selective herbicide which is incorporated into fully tilled soil.

Site-preparation technologies based on these two chemicals were subsequently developed. EPTC was found to have advantages on sites which could be tilled while glyphosate was superior on stony, wet, dry, or sandy sites (Barkley 1983a).

However, field experiences soon demonstrated that EPTC was difficult to apply properly. The timing of EPTC application is critical because the soil moisture content must not be too wet or too dry in order to activate the chemical to produce a gas in the soil which kills all living vegetation contacted in the soil. Moreover, thorough soil tillage is necessary and intact pieces of sod or clods cannot be greater than 10 cm in size (Barkley 1983). Finally, although good site-preparation results could be obtained with EPTC on sites which were recently in agriculture, its use was discontinued in 1984 because sites in the Domtar woodshed were generally too wet or too stony for effective application.

Glyphosate is a more cost-effective herbicide, easier to apply and with a relatively shorter half-life. With glyphosate application, the only critical point is the leaf stage of the target vegetation, otherwise herbaceous vegetation must be mowed first and allowed to resprout.

Tending prescriptions have also been simplified since the early days of the program. Approximately 18 herbicides were tested for weed control efficacy and crop tolerance (Raitanen et al. 1983), but current federal legislation allows only the use of one herbicide – simazine. Previously, it was possible to make herbicide prescriptions based on the weed spectrum present, for example, if triazine-resistant weeds were expected, then chemicals such as terbacil or linuron would be prescribed. Unfortunately, forestry comprises a small percentage of herbicide use in Canada and hence herbicide manufacturers are not interested in pursuing a minor use extension for other herbicides.

The one thing that has not changed, though, is the importance of weed control standards until plantations are free to grow. The challenge for the future will be to develop effective nonherbicide tending options that meet increasingly stringent environmental legislation.

3 Future Perspectives

The hybrid poplar clonal forestry program in Ontario will continue to be refined to meet future requirements. Apart from utilization needs by traditional wood industries, the 1990s will see new and diverse applications for hybrid poplar. Heightened awareness of the greenhouse effect is spurring interest in fast-growing hybrid poplar for carbon sequestering. Environmental concerns regarding run-off pollution and soil erosion on riparian sites is escalating. Global interest in biomass energy is also creating an increased demand for fast-growing hybrid poplar energy plantations. The hybrid poplar clonal forestry program in Ontario will certainly play a strong role as these new initiatives materialize.

To meet future demands for hybrid poplar plantings, specific research and development work must continue in order to strengthen the current clonal plantation technologies. For example, hybrid poplar breeding and selection efforts must continue to add new and improved clones to the production list. Emphasis will be placed on increasing the gene pool material available for breeding, development of early tests for disease resistance screening, especially for *Septoria* canker, and greater inclusion of wood quality criteria in the clonal selection and deployment program.

Silvicultural technologies have been largely developed and refined, but a number of areas still require further exploration, including operational clonal and site-specific fertilization standards, coppice management guidelines, and low or nonherbicide vegetation management techniques. Moreover, we would like to establish optimum spacing and economic rotation production systems which we anticipate may vary for different clones and sites.

The technologies developed to date were derived from a concerted effort to establish, measure, analyze, and interpret results of field trials. Greater use of the data collected will be made in the future. This will include the development of expert systems to assist in field decision-making. In particular, an expert system is being developed to allocate clonal stock to production plantation sites.

The hybrid poplar clonal forestry program has served as a model for the development of clonal forestry programs with other species in Ontario and elsewhere. For example, the FGF group is currently developing a clonal Norway spruce program. Other programs can learn from our experiences. The following three points summarize the important elements to be considered when launching a clonal forestry program.

1. The organization and implementation of clonal programs must rely on stimulating and utilizing applied research, technology development, and operational testing. This process is a continuum, that is, technology development does not begin when research ends.
2. Emphasis must be placed on the breeding, selection and testing of genetic material before major gains can be achieved. As well, intensive cultural practices are imperative to the successful deployment of improved clonal stock.
3. All the partners in a clonal program must have a vested interest in working towards the successful evolution of an operational program. Only then will innovation be realized at all levels of the clonal program, from research to field operations.

References

Barkley BA (1983) A silvicultural guide for hybrid poplar in Ontario. OMNR, Toronto, 35 pp

Barkley BA, Evers RW, Raitanen WE (1983a) Soils: the forest foundation. In: Armson K (ed) New forests in eastern Ontario. Science and technology series, vol 1. Queen's Printer for Ontario, Toronto, p 63–82

Barkley BA, Evers RW, Raitanen WE (1983b) Genes and diversity. In: Armson K (ed) New forests in eastern Ontario. Science and technology series, vol 1. Queen's Printer for Ontario, Toronto, p 83–106.

Evers RW, Morse SH, Rogers DL (1983) Soils: the foundation for poplar plantations in Ontario. In: Stone E (ed) Forest soils and treatment impacts, Proc 6th NA For Soils Conf, Knoxville, Tenessee, p 29

Evers RW, Keane RK, Bowyer LL (1984) Soil survey standards. In: Barkley BA (ed) Soils: the foundation of new forests in Ontario, OMNR, Brockville, p 39–56

Libby WJ (1987) Testing for clonal forestry. Ann For 13:69–75

Muhle Larsen C (1970) Recent advances in poplar breeding. Rev For Res 30:1–67

Raitanen WE, Barkley BA, Evers RW (1983) Setting out the crop. In: Armson K (ed) New Forests in eastern Ontario. Science and technology series, vol 1. Queen's Printer for Ontario, Toronto, p 133–186

Rogers DL, Morse-Krannitz SH, Barkley BA (1984) Soils and clonal growth. In: Barkley BA (ed) Soils: the foundation of new forests in Ontario. OMNR, Brockville, p 23–36

Steenackers V, Strobl S (1988) Short rotation forest biomass plantations in Belgium. In: Hummel FC, Palz W, Grassi G (eds) Biomass forestry in Europe: a strategy for the future. Elsevier, Amsterdam, London, p 273–286

Steenackers V, Strobl S, Steenackers M (1990) Collection and distribution of poplar species, hybrids and clones. In: Mitchell CP, Zsuffa L, Anderson S, Stevens DJ (eds) Forestry, forest biomass, and biomass conversion: the IEA Bioenergy Agreement (1986–1989) Summary Reports. Elsevier, London, p. 1–22

Strobl S (1988) A decade of clone-site testing in eastern Ontario. In: Barkley BA (ed) Poplar culture to the year 2000, Proc Poplar Councus and Can Joint Meet Cornwall, Ontario, p 29–39

Strobl S, Fraser K (1989) Incidence of *Septoria* canker of hybrid poplars in eastern Ontario. Can Plant Disease Surv 69(2):109–112

Tessier du Cros E (1984) Breeding strategies with poplars in Europe. For Ecol Manag 8: 23–39

Wayman M (1978) Wood-fired electricity generation in eastern Ontario. R Commiss Elect Power Plann, 129 pp

Zsuffa L (1979) The breeding of *Populus* species in Ontario. In: Scarratt JB (ed) COJFRC Tree improvement Symp, Toronto, Ontario, p 153–160

13 Policies, Regulations, and Laws Affecting Clonal Forestry

H.J. Muhs[1]

1 Introduction

The task on regulation of forest reproductive material was initiated around 1925 in Germany and other Central European Countries. By that time the international seed trade was expanding and it was realized that there should be regulation to protect forest regeneration from unsuitable seed sources. In 1957, the Federal Republic of Germany amended the rules in which the minimum requirement for approval of basic material, including sealing and labeling of reproductive material for trade and marketing, were prescribed. The amendment contained a very important step in modern legislation concerning the regeneration of forests. While the early rules had aimed at regulating the deployment of forest seed and plants, especially the choice of forest tree species and their provenances for a given site type, the law amended in 1957 sets up rules for marketing forest reproductive material. The reasons for this amendment were strong property rights (the owner of a forest can make his own choice of tree species and provenance) and the right of a free trade. Both rights can to a certain extent be restricted by social obligations. Therefore the law on forest reproductive material has the functions mainly of protecting the consumers' interests, supporting forestry by improving the availability of suitable forest reproductive material, and warranting free trade.

These principles were later adopted and elaborated by the Organization for Economic Cooperation and Development Scheme (OECD, 1974), and the European Economic Community Directives (EEC, 1966, 1969, 1975), for international trade of forestry reproductive material. Up to the present time only two countries, the Federal Republic of Germany (1979) and Sweden (1982), have set up rules and regulations for marketing of clones and clonal mixtures for genera and species other than poplars (Muhs 1982, 1986a, b). In addition, Denmark and Belgium have proposed, but not enacted, regulations on vegetative reproductive materials, while New Zealand and Ontario/Canada have put forth draft proposals, to be initially followed on a voluntary basis, of which the draft proposal from Ontario/Canada is mainly aimed at deployment rather than marketing of reproductive material.

Although the rules for the release of clones and clonal mixtures are most relevant to clonal forestry, there are some more regulations which need to be

[1] Institute of Forest Genetics and Forest Tree Breeding, Federal Research Centre for Forestry and Forest Products, Sieker Landstrasse 2, W-2070 Grosshansdorf, FRG

Clonal Forestry II, Conservation and Application
Ed. by M.R. Ahuja and W.J. Libby

mentioned, namely the UPOV Convention, the patent laws, and the regulation on the release of genetically engineered plant material. But their full impact on clonal forestry cannot be considered in the following discussion.

2 Monoclonal and Multiclonal Plantations

It appears that multiclonal mixtures, also called multiclonal or clonal varieties, will be used in clonal forestry more often than single clones. This is true for most species, except poplars, which are usually planted in monoclonal plantations; however, in some cases they are also grown in multiclonal plantation. For poplars, no regulations concerning the number of clones to be used have been laid down so far, and, therefore, both monoclonal and multiclonal plantations will be accepted. Poplars are an exceptional case, because they have been grown monoclonally for more than 100 years already and their rotation period is rather short compared to most other tree species. The lawyers recognized the ambivalent situation, and therefore, allowed a kind of prescriptive right in the case of poplars. On the other hand, in most other species, and in particular in conifers, multiclonal mixtures are preferred.

Although some estimates on the safe number of clones in a clonal mixture have been presented (Libby 1982, 1987; Krusche 1982; Hühn 1987) the optimal numbers of clones in a clonal mixture are still under discussion for establishing international legal guidelines.

Current theoretical and experimental research in clonal forestry have yielded three general guidelines (Libby 1988): (a) a mixture of large numbers of clones are about as safe as a similar mixture of genetically diverse seedlings; (b) mixtures or regional deployment of very small numbers of clones is not safe, and committment to two to four clones is often worse than monoclonal plantations; and (c) regional deployment of modest numbers (ranging from 7–99 clones) is about as safe as deployment of large number of clones, and offers substantial advantage as well (Libby and Rauter 1984).

In the following discussion the concept of growing a certain number of clones in mosaic designs instead of multiclonal mixtures is not considered, because this has not been a subject either in international or national regulation. In fact, from the legal point of view, mosaic designs are much more like monoclonal plantations than multiclonal ones. Each clone forms a monoclonal plot on a certain area within the mosaic-designed plantation, each clone being identifiable after planting, whereas the clones of a multiclonal mixture are completely randomized and are not easily identifiable any longer after planting.

Besides clones and clonal mixtures, also bulk-propagated plant material is used in clonal forestry. Up to now only Sweden and Ontario/Canada (see below) have included bulk propagated material in their regulation on clonal forestry. The term bulk propagation or bulking of plant material as used by Muhs (1988) usually combines both types of reproduction: (1) sexually produced material (for instance seed) is collected from the basic material to be propagated and grown to young plants, and (2) the plants derived from the seed are multiplied by vegetative means (for instance cuttings). The main feature of bulk propagation or bulking is that the identity of the plants multiplied is lost during multiplication. In the Swedish regulation the term bulk propagation is defined differently: as basic material a seedling or an "unidentified clone" can be used, which is multiplied by vegetative means; in this case the first (sexual) step of propagation is not important. The result in all cases, however, leads to plant material in which single clones cannot be identified.

3 Marketing of Forest Reproductive Material

The marketing of reproductive material is governed by international rules such as the OECD-Scheme for the Control of Forest Reproductive Material Moving in International Trade (1974) and the EEC Council Directives on Marketing of Forest Reproductive Material (1966, 1969, 1975) or by national law, which, as in Germany or other EEC member countries, is based on the above mentioned EEC Directives, whereas Sweden follows its own rules, and Ontario/Canada and New Zealand are governed by recommendations set up for national or regional application. In this context, the international rules are of interest, and can be characterized by the following three main principles: approval, identification, and control.

3.1 The Principle of Approval

The principle of approval is the basis of marketing reproductive material. The basic material, which may be a seed stand, a seed orchard, a clone, or clonal mixture, is the object of an official act, called approval.

The act approval gives permission to an owner of the basic material to collect reproductive material from the basic material and sell it in the market. It is generally accepted that from seed stands or seed orchards only seeds or seedlings should be collected as reproductive material. The law does not permit the marketing of other reproductive materials, such as vegetative material to produce rooted cuttings from a seed stand or a seed orchard. Similar rules apply to clones and clonal mixture. The law permits the marketing of vegetatively propagated material or plants derived from an approved clone or clonal mixture. However, it is not permitted to market seeds or seedlings originating from the approved clones or clonal mixture.

Thus the principle of approval is based on the following premises:

- The physical existence of basic material is: a seed stand of a given size, or a seed-orchard of a given composition or a specimen of a clone in a clone bank or a given set of specimens of clones for producing clonal mixtures.
- The mode of reproduction is: either sexual or asexual (vegetative).

The terms reproductive material and reproduction are used in the following context:

- the reproductive material comprises seed and vegetative parts;
- sexually derived plants in the case of seed stands and seed orchards (sexual propagation);
- asexually derived plants in the case of clones or clonal mixtures (vegetative propagation).

The terms do not distinguish between the reproduction and the multiplication of reproduced material in the strict sense. Thereafter, reproduction of a parent tree means that the seeds derived have undergone meiosis followed by fertilization, while reproduction of an ortet results in ramets, which have been produced via mitotic cell divisions only, without undergoing reduction division (meiosis).

The principle of approval describes the relationship between the terms basic material and reproductive material as follows: the basic material gets the approval, while the reproductive material derived from the basic material goes on the market (Muhs 1988). The approved basic material has to be listed in a document containing information about category, Latin name, identification, location, origin, ecological conditions, seed-orchard, and test results as applicable, which is outlined more precisely in the OECD Scheme (1974) or in the EEC Directive (1966) respectively.

The act of approval is therefore a most important one, because it gives the possibility to clearly define the pedigree of the reproductive material and to assess judicial revisions in the case of contestation of a consignment of reproductive material.

While the OECD Scheme (1974), the EEC Directives (1966, 1969, 1975), and the national laws of the EEC member countries have adopted the approval obligatorily, other countries either did not, or they specified a less stringent form of approval called permission. This, for example, is the case in the Swedish law on regulations in clonal forestry (1991). Here it is said that "for trade with cuttings a permit is needed." Further specifications are lacking (which does not mean that there are none, but that they have not been specified by law).

Examples for regulations without approval of the basic material are those from Ontario/Canada and New Zealand. In Ontario/Canada the draft outlines the requirements to be followed when allocating tree improvement stock to operational planting projects. The aim of this regulation, which has been proposed in the draft, is to intensify tree improvement activities in order to produce artificial regeneration stock with greater genetic potential. It will ensure a broad genetic base in its new forests by regulating the deployment of all tree-improvement stock. Instead of governing the trade with forest reproductive material, especially tree improvement stock, the deployment will be regulated. There is no act of approval foreseen for the tree improvement stock. To make sure that the regulation will work in practice, the District/FMA Holder, the Region, and the Forest Resources Branch have to take over management responsibilities for a total of 13 activities described in the above-mentioned procedure.

Another example is the Certification System for Forest Tree Seed and Planting Stock from New Zealand (Vincent 1987): The scheme operates on a voluntary basis. The Seed Certification Service is run by the Ministry of Forestry's Genetics and Tree Improvement Group at the Forest Research

Institute in Rotorua on behalf of the New Zealand forest industry. The scheme gives advice in seed classification, plant classification, certification, certification procedure, and the seed register. There is no act of approval of basic material for the production of seed or planting stock described. The tree seed and plant certification service is intended to assist users to identify more easily the characteristics and genetic quality of seed and plants, and to certify that the seed plants are true to label. The label or packing note remains, however, the suppliers' personal guarantee as to the identity of the seed or plants (cited by Vincent 1987). The scheme is simple and easily applicable and seems to work on a voluntary basis without any official acts on approval, identification, and control.

3.2 The Principle of Identification

This principle is based on a reference number or letter, which is given to the approved basic material. By this the basic material gets an identity, which enables the Designated Authorities responsible for the implementation of the scheme to follow directives to certify reproductive material derived from a certain approved basic material. This principle has been adopted in the OECD Scheme and in the EEC Directives and in the national laws of the EEC member countries. Each approved basic material has to be listed in a national list as mentioned above using the reference number or letter. A code system for the reference numbers or letters is outlined in the scheme or directives. This is also true for clonal stock. That is, the ortet of a clone is the basic material from which ramets are derived by vegetative propagation. The ortet gets approval and a reference number for identification and the location where it grows has to be described. Instead of the ortet, a specimen (ramet) derived from that ortet and planted in a clonal bank can serve as basic material for the official act of approval and identification. Only reproductive material derived from the approved ortet or the approved specimen is allowed to be marketed under the name of the clone.

The identification is a logical part of the approval, that is approval without identification is incomplete, because only approved and identified basic material can be controlled. In the New Zealand scheme (Vincent 1987), the certification involves evaluation by application forms, classification according to the system described, and registration of the seedlots certified, but does not involve their identification, whereas the draft of the tree-improvement stock deployment in Ontario/Canada (1988) says that the District Management/FMA Holder has to keep "detailed records of planting block sizes and locations, numbers of clones, and clone identifications." Their is no further information given, so that it cannot be compared with what has been said above.

3.3 The Principle of Control

It is obvious that rules without control systems are weak. The reliance on the rules depends on the agencies involved and their responsibilities. As already mentioned, the examples for a scheme as presented by New Zealand and Ontario/Canada do not set up a control system, while the OECD Scheme, the EEC Directives, and the national laws of the EEC member countries contains a more or less extensive control system. The system consists of two parts: a label accompanying the lots of reproductive material from the place of collection/production via all steps of processing, storing, transporting, raising, and marketing to the place of final use, and a controlling agency, which is independent and/or public, in most cases under the responsibility of the designated authority. The labels contain information on the following (as applicable):

- name and address of the designated authority,
- species (Latin name),
- subspecies, variety, or cultivar name,
- category,
- reference number of the Certificate of Provenance or Clonal Identity,
- provenance or region of provenance.

The labels are colored according to the following categories (after the OECD Scheme):

- Source identified reproductive material yellow
- Selected reproductive material green
- Untested seed orchard material pink
- Tested reproductive material blue

Only the categories "selected reproductive material" and "tested reproductive material" apply to the EEC Directives. Clonal material can be marketed with green (selected reproductive material) or blue (tested reproductive material) label after the OECD Scheme and EEC Directives. However, some countries (for instance the Federal Republic of Germany) permit the marketing only as tested reproductive material depending on species.

The labeling of the lots of reproductive material allows the complete and consistent control of the identity of a lot of reproductive material. For both international rules, the OECD Scheme and the EEC Directives, the control is mandatory.

From the above discussion the following conclusions can be drawn. The application of the principles of approval, identification, and control in the international or national rules gives evidence to a high degree of protection of the consumer's interests. On the other hand, it complicates the rules to such an extent that some countries do not adopt the international schemes and try to find simpler ways, which do not comply with the international rules. As long as there is no trade with countries that are not governed by the international schemes, there are no problems. Thus, those countries are isolated from the international market and have to accept this. Further, it is evident that the act of approval has highest impact on breeding clonal varieties.

4 Laws and Regulations for Clones and Clonal Mixtures

While both international schemes, the OECD Scheme and the EEC Directives, have only general rules, which include vegetatively propagated reproductive material suitable in the categories of the schemes, the marketing of reproductive material of clones generally is not prohibited. However, because of lack of specified regulations, they are not prepared to give guidance for the marketing (approval and labeling) of clonal material in mixtures. Only Sweden and the Federal Republic of Germany have set up laws, which are compared in Table 1. The draft from Ontario/Canada contains a proposal which has also been included in Table 1. Here 15 items of general interest are listed for comparison, showing the differences between the regulations. Sweden has amended its rules recently (February 1, 1991), which is considered here, but not in the synopsis published by Muhs (1986a, 1988) earlier. The rules from New Zealand are not included in Table 1 because they do not have specific regulations for clonal mixtures, are too general, and have application only for classifying cuttings.

Table 1 shows that the approaches adopted by the lawyers of the three countries for setting up rules for clonal mixtures are different. Sweden has the oldest rules, which have been prepared for regulation since 1982. They apply both to practical use and commercial and noncommercial transfer of coniferous cuttings. The Federal Republic of Germany had enacted the law in 1979, and details were set up in the General Administrative Regulation from 1985, which can be applied to forest reproductive material of all species. The task is to govern the marketing of clonal mixtures, and not to regulate the practical use. This is just contrary to the draft from Ontario/Canada, which aims at regulating the deployment of the clonal stock.

Beside the 15 items considered in the Table 1, there are more subjects, which have been laid down in the rules. In some cases they regulate highly specific details, which are not of interest in this context. For more details on laws and regulations of forest reproductive material see references in Table 2.

5 The Approval of In Vitro-Propagated Plant Material

Recent advances in tissue culture technology have paved the way for rapid clonal propagation of woody plants. Regeneration of plants in vitro (micropropagation) may be accomplished by organ, callus, cell, or protoplast cultures. Clonal fidelity of in vitro-regenerated plants is important in clonal forestry programs. By employing tissue culture technology, true-to-type plants are not always produced. By and large, plants produced by organ culture (for example bud meristems) are relatively stable and involve minimal genetic risk. On the other hand, long-term callus cultures, derived from cells or protoplasts, are generally prone to genetic variability. Genetic variation originating in vitro (termed somaclonal variation) would be useful for recovering desirable genotypes, for example disease resistance. Plants produced following genetic engineering, for example gene transfers involving foreign genes, also may or may not

Table 1. A synopsis of rules governing the marketing or the deployment of clonal material

Subject	Sweden	FRG	Ontario/Canada
1. Legal status of the rules	Law since 1982 amended in 1991	Law since 1979, details are laid down in a General Administrative Regulation amended in 1985	Not yet enforced
2. Selection of basic material	Not regulated	Not regulated	Not specified
3. Multiclonal mixtures, mosaics, or single clones for species other than poplars	Multiclonal mixtures are recommended; row planting is preferred over block planting if clones are kept identifiable	Multiclonal mixtures only	Random or sequenced mixtures, clonal blocks
4. Species covered other than poplars	Conifers	All species (18 totally under the law, for the genus *Populus* single clones and mixtures are allowed)	Not specified, seems to apply for all species including hybrid poplars
5. Minimum number of clones per mixture	According to test levels, the percentage for a single clone in a plantation should not exceed for: untested clones 1.5% (not less than 67 clones) Clones tested at level 1 2.5% (not less than 40 clones) Clones tested at level 2 3.5% (not less than 29 clones)	Dependent on species; 500 (100) for main species used for large area reforestation; 100 (20) for less important species used for small area forestation. The number in brackets is the minimum number for special sites, for which special clones have been tested	A planting block shall contain 15 open-pollinated or unrelated families in equal proportions for improved seedling based stock or bulk-propagated clonal stock. If some families occur substantially more frequently than others, the minimum number to be 20 families; 100 tested clones in the production set for improved clonal stock. For each planting block the minimum number of clones that are to be planted are (y.e.r.a. = years of expected rotation age) Size (ha) — Minimum number of clones (y.e.r.a $<$ 25 / y.e.r.a. $>$ 25) up to 20: 7 / 10 21 to 50: 15 / 20 50 +: 25 / 40

6. Description of the identity of clones	Not specified	Description of each clone according to provenance or origin or, if it is a breeding product, descent and some additional items	Not specified
7. Use of untested clones	Yes	Not allowed	Yes
8. Testing procedure:			
– Single clone or mixture testing	Single clone	Single clone	Single clone
– Duration of test	According to test level level 1: 6 years level 2: 9 years and additionally two replications at level 1	Dependent on species from about 10–30 years commonly used in progeny testing, (minimum number of years for testing not given)	1/4 of rotation age
– Number of test sites	Level 1: 2 sites level 2: 4 sites	At least 3 sites	Not specified
– Traits to be analysed	Climatic suitability, growth vigor, survival, damages, and additional traits if desired	Identification, adaptation, growth vigor, resistance	Not specified
– Control (standard)	Seedlings suitable for the test area shall be included for comparison	Detailed specification for each species in a list of "Standards"	Not specified
– Evaluation of results	Not specified	Procedure has to follow special rules (the superiority must be significant at the 95% level)	Not specified
9. Bulk propagation	Not more than 200 ramets from each plant derived from seed or from a not identified clone may be bulked up	Not allowed	Bulk propagation is allowed, untested clones to be treated as bulked material
10. Mixing procedure, clonal identity	Not specified	Clone identity is maintained in all the following propagation cycles and lost only when material is combined in mixtures for use in forestry	Not specified

Table 1 continued

Subject	Sweden	FRG	Ontario/Canada
11. Restrictions for use	Suitable climatic zones for the material must be declared	Mixing before use, if tested for special sites, this shall be specified on the label	Contiguous planting blocks planted as separate projects, or planted over a period of up to 5 years, to be considered as one planting block and clones allocated according to the sum total of all blocks. Planting blocks up to 1000 m apart to be considered contiguous blocks for the purpose of clone allocation
12. Approval	By the National Board of Forestry	By the authorities of the Federal Lands as "tested reproductive material"	Not specified
13. Limitation of approval	According to test level, untested: 50 000 level 1: 250 000 level 2: 700 000 ramets per clone	10 years, prolongations for further 10 years possible	Not specified
14. Registration of clones	Not specified	To be done by the responsible authority	Not specified
15. Control	Not specified	To be done by the responsible authority	Not specified, while the Forest Service at the District, Region, and Forest Resources Branch level is involved in a number of actions, there are responsibilities and thus indirect control

Table 2. Rules and regulations for forest tree reproductive material

OECD (1974): OECD Scheme for the Control of Forest Reproductive Material Moving in International Trade. Organization for Economic Co-Operation and Development (OECD), Directorate for Agriculture and Food, Paris, 24 pp.

EEC Directive (1966): Council Directive of 14 June 1966 on the marketing of forest reproductive material (66/404/EEC), Official Journal of the EC No. 125, 11.7.66, pp. 2326/66.

EEC Directive (1969): Council Directive of 18 February 1969 amending the Council Directive of 14 June 1966 on the marketing of forest reproductive material (66/64/EEC), Official Journal of the EC No. L 48, 26.2.69, p. 12.

EEC-Directive (1975): Second Council Directive of 26 June 1975 amending Directive No. 66/404/EEC. Official Journal of the EC No. L 196, 26.7.75, pp 14–23.

German law on the marketing of forest reproductive material (1979): Gesetz über Forstliches Saat- und Pflanzgut, Neufassung vom 26. Juli 1979, Bundesgesetzblatt, Jahrgang 1979, Teil I, pp. 1242–1261

Swedish law on the marketing of forest reproductive material, amended 1 February 1991: regulations on Clonal Forestry, Statute Book SKSFS 1991 : 1, 4 pp., English translation available from the National Board of Forestry, Jönköping, Sweden.

New Zealand's regulations (1987): See Vincent (1987)

Ontario/Canada's draft for a regulation (1988): Tree Improvement Stock Deployment. Draft No. 2, Procedure FR 12 04 01 and FR 12 04 10, January 1988, Ministry of Natural Resources, Genetic Resources Management, Ontario/Canada.

UPOV Convention (1961): International Convention for the Protection of New Varieties of Plants of December 2, 1961, and Additional Act of November 10, 1972, and Revised Text of October 23, 1978. International Union for the Protection of New Varieties of Plants (UPOV), 293 (E), Geneva, 1985, pp. 64.

be genetically stable. These are some of the problems that are facing the law makers in Germany and the European Economic Community for release of biotechnologically produced plant material in the international trade.

Once the tissue culture and genetic engineering techniques are standardized for production of relatively stable genotypes, then one could handle the micropropagation of selected material in the same way as vegetative propagation of cuttings (Muhs 1988). As in the case of bulking, micropropagation from unselected seed via cotyledon or embryo culture may not be looked upon favorably by the law makers. Therefore, in this case the basic material must fulfill the requirements to gain approval, as outlined in the previous section. The material(s) must be stable and true to type. It should be tested in comparison to one or several standards before getting the approval; the conditions of the test are laid down in the rules. They should be marketed as clonal mixtures according to the German rule, if species other than poplars are in question. However, there might be some problems in finding standards for newer genotypes produced by genetic engineering.

With regard to genetically engineered plant material, the United States Department of Agriculture (USDA) has set up proposals for the introduction of field testing (Purchase and MacKenzie 1990) in the United States of America. In 1990, the Federal Republic of Germany enacted a law on the release of genetically engineered and biotechnologically produced plant material, while similar proposals on conducting safe experiments outside the laboratory on genetically engineered microorganisms and plant materials are under discussion in Europe, and by 1992 an EEC Directive on research on genetically modified organisms is expected to be released. The question of special interest is: does forest plant material, which gained permission for release to the market according to the rules for the release of genetically modified organisms, automatically have the approval or permit for use in practical forestry? The logical answer should be: no, it should not, because a new forest plant variety, which has been safely produced by genetic engineering and is released to the market, does not necessarily meet the requirements laid down in the rules for the marketing of forest reproductive material as shown above. It has to be tested as a clone or in multiclonal mixture first before marketing. From this it follows that genetically modified forest plant material has to pass two rules and meet the requirements of both rules, first for the release of genetically modified organisms, and second for the marketing of forest reproductive material.

6 Concluding Statement

It has been shown in the previous sections that regulations and laws for forest reproductive material are affecting clonal forestry to a high degree. Principally, there are two different approaches when comparing the regulations: guidelines ruling the deployment of clonal planting stock (for example Ontario/Canada), and regulations governing the approval of basic material and the marketing of reproductive material derived from the basic material. In the second case an identification for each basic material is necessary, which is part of the act of approval and gives the possibility of establishing a control system. While the first approval ruling the deployment is of only regional importance, the second is widely adopted, including both international regulations (OECD Scheme and EEC Directives). There seems to be a trend towards following the international rules. This will be more attractive when the OECD and the EEC have included special rules for clonal varieties, which are under discussion now. The approval involving biotechnologically produced or genetically modified plants, as well as plants derived vegetatively from untested breeding stock by bulk propagation, where the breeding stage is advanced, but testing not completed because of the long test duration, is difficult or not possible according to the existing regulations. The OECD and the EEC are making efforts to amend their regulations in order to update them and to prepare the prerequisities for marketing these plant materials. A new category is under discussion, which is in between the category "selected" and "tested" reproductive material. This category could be entitled "genetically improved", for which the genetic improvement has to be ensured by calculations of the expected genetic gain. That is, the expectation of a genetic gain is not enough, it has to be demonstrated by data which show evidence of a genetic gain.

The OECD Scheme should take care of its international role for marketing forest reproductive material. On the one hand, it has to consider developments in breeding and propagation techniques and thus being attractive for countries with advanced breeding programs. On the other hand, it should be as simple as possible to give guidance to all countries, which have not implemented a scheme so far. This is important for third world countries especially, which are going to afforest large areas using planting stock of unknown quality or which are able to provide the market with planting stock of known origin. In both cases the implementation of the OECD Scheme protects the interests of consumers, producers, and traders.

References

Hühn M (1987) Clonal mixtures, juvenile-mature correlations and necessary number of clones. Silv Genet 36:83–92

Krusche D (1982) Host–parasite interactions and the number of clones in a clonal mixture. In: Proc IUFRO Joint Meet Working Parties Genetics about breedings strategies including multi-clonal varieties, Sensenstein, p 171

Libby WJ (1982) What is a safe number of clones per plantation? In: Heybrook HM, Stephan BR, von Weissenberg K (eds) Resistance to diseases and pests in forest trees. PUDOC Wageningen, pp 342–360

Libby WJ (1987) Testing for clonal forestry. Ann For 13:69–75

Libby WJ (1988) Testing and deployment of brave new plantlings. In: Ahuja MR (ed) Somatic cell genetics of woody plants. Kluwer, Dordrecht, pp 201–209

Libby WJ, Rauter RM (1984) Advantages of clonal forestry. For Chron 60:145–149

Muhs HJ (1982) Identifying problems and legal aspects of using multiclonal mixtures. In: Proc IUFRO Joint Meet Working Parties Genetics about breeding strategies including Multi-clonal varieties. Sensenstein, FRG, pp 92–108

Muhs HJ (1986a) A synopsis of some international and national rules for forest reproductive material moving into trade. OECD Scheme, EEC Directives, and German law of forest reproductive material. Mitt Bundesforschungsanst Forst Holzwirtsch (Hamburg) 154:25–37

Muhs HJ (1986b) A synopsis of some national regulations for multiclonal mixtures as applied to Norway spruce and species other than poplars. Mitt Bundesforschungsanst Forst Holzwirtsch (Hamburg) 154:202–209

Muhs HJ (1988) Rules governing the release of forest reproductive material derived by in vitro culture. In: Ahuja MR (ed) Somatic cell genetics of woody plants. Kluwer, Dordrecht, pp 211–219

Purchase HG, MacKenzie DR (eds) (1990) Agricultural biotechnology, introduction to field testing. Off Agric Biotechn USDA, 58 pp

Vincent TG (1987) Certification system for forest tree seed and planting stock. Minist, For Res Inst, Rotorua, NZ, FRI Bull 134: 17 pp

Glossary

The **bold-face numbers** following the words and phrases in this Glossary refer to chapters in which the word or phrase is used in context, or in which it is more fully defined. This Glossary was prepared by the Editors, with consultation and help by the Chapter Authors, and by Dr. P. Mathias and Dr. S. Jelaska, Visiting Scholars in Berkeley during 1988–89 and 1990–91. We have drawn on Glossaries (Snyder 1959; Futuyma 1986; Pierik 1987; Curtis and Barnes 1989; Lackie and Dow 1989; Rieger et al. 1991; Donnelly and Vidaver 1988), descriptions in context (Grierson and Covey 1988), and on a dictionary (Guralnik 1976), frequently modifying definitions to better fit the context of this volume.

Actinomorphic: **10** Flowers capable of bisection into similar halves in two or more planes.

Additive genetic variation: **6**. The proportion of genetic variation that responds to natural selection, mass selection, or pick-the-winner selection. This is the basis of a parent's breeding value, or general–combining–ability (GCA).

Adventitious: **2**, **5**, **9**. Meristems developing into buds, shoots or roots in places not directly associated with apical meristems or out of the normal phyllotactic pattern of the plant.

Anemophilous: **4**. Wind pollinated.

Anthraquinones: **10**. A class of organic compounds often found in plants as pigments.

Architecture: **3**. See genetic architecture.

Archives: **3**. For genetic conservation: Ex-situ collections of germplasm.

Basic material: **13**. Material from which the (reproductive) production material is obtained. See foundation stock (Vol. 1) expansion stock.

Biotype: **9**. Two biotypes are the two sets of phenotypes that develop in two clones.

Budding: **2**, **10**. A specialized form of grafting, in which a scion bud and adjacent cambial tissue is slipped below the bark of the stock.

Bulking up: **2**, **11**, **13**. See vegetative multiplication.

Chronological age: **6**. Years since germination of the ortet. Contrast to maturation state, which may vary with location in the tree as well as with ramet history.

Coppice: **5**, **9**. New shoots from the stump or roots. To coppice is to cut the main stem at the base or to injure the roots in order to utilize coppicing for regeneration.

Cryopreservation: **3**. Storage of cells, tissues, seeds etc. at extremely low temperatures, usually in liquid nitrogen (-196° C).

Culm: **10**. The stem of grasses and bamboos; usually hollow except at the nodes.

Cultivar: **2**, **4**, **5**. A 'cultivated variety', given a non-Latin name and designated by 'cv'. Any clone, race or product of breeding deemed worthy of a separate name.

Cutting: **11**. A shoot, twig or other plant part removed from the donor plant, usually for the purpose of inducing roots to produce a steckling.

Cyclophysis: **4**, **11**. The maturation of apical meristems may lead to differences in performance among vegetative propagules taken at different times from members of the same clone or even from the same donor plant. See periphysis (Vol. 1) topophysis.

Demographic: **3**. Relating to density, age, and distribution of individuals in a population.

Deployment: **6**, **7**, **9**, **12**, **13**. The physical movement of clones or other genetic units from one site (usually a nursery) to another (usually plantations), including their locational configuration on the recipient site.

Desoxylapachol: **10**. A toxic fungicidal organic compound belonging to the class of substituted napthaquinones.

Dysgenic: **3**. Causing a reduction of desirable genetic qualities in natural or production populations.

Embryogenesis: **10**. Process by which an embryo initiates and develops from a zygote or, asexually, from a somatic cell or group of cells.

Embryoid: **5**. Embryo-like structure produced by somatic cells in vitro; also, adventitious embryo developing in vitro, in this latter sense = somatic embryo.

Endocarp: **10**. The inner layer of the fruit wall (the pericarp).

Epicormic: **9**. Initiating from the stem, as when dormant or suppressed buds elongate and form epicormic shoots following exposure or fire.

Expansion stock: **8, 11**. Propagule donors that originate from foundation stock (Vol. 1) and that are used to produce large numbers of propagules for production outplantings.

Ex situ: **3**. Off the site; away from the natural habitat.

Extirpation: **3**. Local extinction; loss of some but not all populations of a species.

Family forestry: **8**. Tested open-pollinated, polycross or full-sib families are deployed as single families or as prescribed family mixtures.

Family hedges: **11**. Hedges used to donate cuttings, in which sibs from the same full-sib, polycross or open-pollinated family are mixed, and their cuttings are bulked. Used for vegetative multiplication at the family level.

Farm-field test: **11**. A test done at close spacing under near-ideal conditions, usually for the purpose of early culling.

Fastigiate: **5**. Having a narrow crown with ascending branches.

Full-sib progeny: **6**. The offspring resulting from a cross of two unrelated parents.

Furnish: **7**. Both the quantity and quality of wood supplied to a mill, usually in reference to a pulp mill.

GCA: **6**. General combining ability. See additive genetic variation.

Genetic architecture: **2,3**. The distribution of genetic variation in a species, usually described hierarchically as variation at the regional, local, family and individual levels, and also relating to proportions of additive and non-additive inheritance.

Genetic gain: **6, 7**. The average (heritable) change from one generation to the next as a result of selection.

Germplasm: **8, 10, 11, 3**. Genetic material. See reproductive material.

Graft incompatibility: **2**. A destructive interaction between tissues of the stock and scion, often resulting in starvation and death of the rootstock. See grafting.

Grafting: **2**. A form of cloning in which a bud or twig (called a scion) of the clone is inserted into a slit on a rooted and established stock plant, which may or may not be clonal. Thus, only the scion portion of a graft is usually clonal.

Hardening off: **11**. By exposing propagules to increasingly harsh environmental conditions, they are weaned from the abnormally favorable conditions of the nursery, greenhouse, or culture room, and become ready for field planting.

Hedge-Orchard: **6, 9, 11**. An orchard of hedged plants that produces cuttings or other tissue for vegetative propagation. See hedging.

Hedging: **2, 6, 8, 9, 11**. The repeated clipping of an ortet and/or some of its ramets to produce a low hedged plant. The purposes of this procedure are to slow or halt maturation of the plant, and to control its size and shape for efficient operation.

Hermaphrodite: **10**. Bisexual; more narrowly, a plant with male and female organs in the same floral structure.

Heterosis: **5, 7**. The average of the progeny is better than the average of the parents. Sometimes called hybrid vigor.

Histogenesis: **5**. The process of formation and development of tissues.

Hybridization: **2, 6**. Interspecific: a cross between species. Intraspecific: a cross between populations within a species, or even between individuals of contrasting genotype within a population. Introgressive: the moving of genes from one species or population to another by repeated backcrosses.

IBA: **11**. Indole-butyric acid, a synthetic auxin used to enhance rooting.

Ideotype: **2**. An idealized multitrait characterization; for example, a crown that is long, narrow and dense with high branch angles.

In vitro: **3**. "In glass"; in aseptic, culture under laboratory conditions. See *in vivo* (Vol. 1).

In situ: **3**. On site; within the natural habitat.

Inbreeding: **9**. A reduction in average heterozygosity resulting from a mating between relatives.

Inoculum: **5**. Pathology: a living organism introduced into a different (host) organism. Tissue culture: the cells or tissue placed on or in a nutrient medium.

Keystone ecological role: **3**. Situation in which the natural presence of a species is central to the viability of many other organisms.

Latent root primordia: **5**. Performed root primordia in the stem with the potential to renew development to form roots when activated.

Maturation: **6, 11**. A sometimes abrupt (heteroblastic) or, more common, unevenly gradual (homoblastic) process of orderly development from embryonic through juvenile and adolescent to mature.

Micropropagation: **13**. The *in-vitro* vegetative propagation of plants, often called plantlets, micropropagules, or somatic embryos.

Mitigation: **3**. Restoration actions taken on a species or population specifically to compensate for previous or (on other sites) planned destructive actions.

Multiclonal variety: **5, 6**. A cultivar (variety) composed of several compatible clones.

Multilines: **2, 6**. Sublines of the general breeding population purposefully selected for different sets of traits or deployment destinations.

Multiple-stage testing: **12**. See sequential testing, combined testing (Vol. 1). May also include separate tests for different purposes, such as resistance to different pests and stresses.

Non-additive genetic variation: **6, 7**. The proportion of genetic variation that doesn't respond to simple mass selection, and that causes specific pairwise crosses to depart from performance values predicted by the breeding values of the parents.

Open-pollinated progeny: **11**. Progeny from a mating that is not controlled. A mixture of related and unrelated pollen is delivered by wind, or in some cases by insects etc. The o-p family is typically a mixture of full-sibs, half-sibs, and selfs, or, in a natural stand, of such relatives with various degrees of inbreeding.

Ortet: **5**. The initial individual (usually from a zygotic embryo) that is vegetatively propagated to produce a clone. See ramet.

Osier: **5**. A shrubby willow (for example, *S. viminalis*, *S. fragilis*) with the ability to develop long slender shoots from a cut base that are suitable for basket production.

Pedigree: **9**. A record of parentage, often including data on performance of the parents and of other relatives.

Photoperiod: **5**. Day length. Used to describe the relationship between phenology and episodes of light.

Plagiotropic: **6, 11**. A branchlike growth form of an independent plant, usually at an angle other than vertical and often bilaterally symmetrical.

Plagiotropism: **2** . A tendency for a propagule to grow at an angle and like a branch, in contrast to vertical, or orthotropic, growth.

Plantlet: **5, 8**. Broadly, a plant produced *in vitro*. Narrowly, the intensive-care stage of a plant produced in tissue-culture. See plantling.

Plantling: **8, 9**. A propagule of tissue-culture origin that is past the intensive-care stage of early nursery care, and can be or has been planted in normal field conditions. See embling (Vol. 1) steckling, hardening off.

Plus-tree: **4, 6, 9**. A tree or genotype selected on the basis of its outstanding single-copy performance, but not yet clonally tested or progeny tested.

Pollard: **5**. Removal of all branches from the trunk, followed by sprouting of new shoots to form a new crown.

Polyculture: **2**. The intimate mixture of several species; in agroforestry, these include animal and/or crop components in addition to one or more tree species or cultivars.

Production clones: **12**. Those clones maintained in large enough stoolbeds, hedges, cultures, etc. to supply large numbers (say 20,000) of cuttings per year for planting. See production population (Vol. 1).

Progenic: **3**. Causing improvement of genetic qualities in natural or production populations of a species. Similar to eugenic.

Progeny test: **6, 11**. Generally a common-garden test in which the breeding values of parents are evaluated and ranked on the basis of the performance of their offspring. See progeny trial, provenance test.

Progeny trial: **6, 11**. A trial in which pedigreed progeny are used either to backward-select among the parents (a progeny test) or to forward- select among the offspring (not a progeny test).

Provenance: **6**. The geographic origin of a population. Mostly (but not always in European usage) the ultimate natural origin, implying where the population evolved prior to human intervention.

Provenance test: **6**. A common-garden test in which population samples from stands of known evolutionary origins are grown together in fair comparisons in one or more locations.

Ramet: **5, 6**. All vegetative propagules of an ortet are ramets. A clone is composed of the ortet and its ramets.

Ramicorn: **9**. A large high-angled branch, often resulting from one member of a fork being partly suppressed by the more dominant member.

Rejuvenation: **9**. A change in a tissue or an organism from a more mature state to a more juvenile one.

Reproductive material: **13**. All material produced by sexual and asexual means used for plant production. See germplasm.

Restoration: **3**. Actions taken to promote genetic or ecologic recovery of a population or species following natural or artificial destruction or impoverishment of the population or species.

Scion: **2**. The (usually desired) clonal plant part, often a twig, that is grafted onto the root-bearing part of another plant. See stock.

Seed orchard: **6, 10, 13**. An orchard consisting of clones or seedlings from selected trees, isolated to prevent or reduce pollination from outside sources, and cultured for early and abundant production of seeds for reforestation.

Seed source: **6**. The geographic source of a seed. If a native stand, equivalent to provenance.

Seed stand: **6, 13**. A plantation or natural stand designated for repeated collection of seeds for reforestation purposes.

Seedlot: **11**. A collection of seeds, usually of known origin.

Sequential testing: **12**. A testing approach in which large numbers of genetic entries are tested in few replications in the early stages (say screening and then candidacy testing) and progressively fewer entries are then tested in more replications or in larger plots (say performance or compatibility testing). See combined testing (Vol. 1), multiple-stage testing.

Serial propagation: **2, 6, 11**. The ortet donates cuttings or tissue that become primary ramets, some of which in turn later donate cuttings or tissue that become secondary ramets, etc.

Somaclonal variation: **2, 3**. Variation arising in tissue-culture of somatic tissue, often at unusually high rates due to higher rates of gene mutation, transpositions, and various chromosomes abnormalities. See gametoclonal variation (Vol. 1).

Somatic embryogenesis: **2**. A process by which somatic cells are differentiated into somatic embryos.

Steckling: **6, 8, 9, 11**. A rooted cutting that is sufficiently developed so that it can be or has been planted in the field. See somatic seedling, (Vol. 1) plantling for parallel terminology.

Stochasticity: **3**. Of or pertaining to chains of random events.

Stock: **2**. For grafting, the rootstock on which the scion is grafted.

Stocktype: **9, 11**. The kind of propagule sent to the field, such as container-grown or bare-root; seedling, steckling or plantling; etc.

Sublines: **2, 6**. A more-or-less arbitrary separation of lines within a breeding population; breeding crosses are made within lines and production crosses between lines, resulting in no inbreeding in the production stock.

Technology development: **12**. The process of developing procedures through trials and experiments. See technology transfer.

Technology transfer: **12**. The stage in development of a technology or process in which operational personnel learn about the technology, commit to it, and develop procedures for operational-scale application.

Testa: **10**. The outer coat of the seed.

Tetralocular: **10**. Consisting of four cavities or compartments (locules) in the ovary or fruit.

Tissue culture: **9**. A general term for aseptic cell, tissue, organ and protoplast culture.

Topophysis: **4**. Differences in maturation or developmental potential among apical meristems of different branch hierarchical order may result in differences in performance among vegetative propagules taken from different parts of the same donor plant. See cyclophysis, periphysis (Vol. 1).

True-to-type: **13**. Individuals having essentially the same phenotype as the donor plant, cultivar or clone.

Variety: **5**. See cultivar.

Vegetative multiplication: **2**, **6**, **8**, **11**. When seeds of desired families or populations are in short supply, embryos or seedlings are clonally reproduced by rooting cuttings, tissue culture, grafting, or somatic embryogenesis. Clonal records are usually not kept.

Whip: **5**. A straight slender shoot with high elasticity. Often used to describe a large unrooted cutting for field planting.

Withe: **5**. A slender flexible branch or twig.

Zygotic forestry: **2**. Whether seedlings or clonal propagules, individual genotypes are not well known or characterized. Performance knowledge is at the level of family, breed or population. See vegetative multiplication.

References

Curtis H, Barnes NS (1989) Glossary pp G1-G22. In: Biology, fifth Edition, Part 2. Biology of organisms. Worth Publ. Inc. New York, 1192 pp.

Donnelly DJ, Vidaver WE (1988) Glossary of plant tissue culture. Belhaven Press, London 141 pp.

Futuyama DJ (1986) Evolutionary biology. 2nd Edition. Sinauer Assoc. Sunderland, Ma. 600 pp.

Grierson D, Covey SN (1988) Plant molecular biology. 2nd Edition. Blackie, London. 233 pp.

Guralnik DB (Ed.) (1976) Webster's New World dictionary of the American language. Second College Edition. Wm. Collins and World Publ. New York, 1692 pp.

Lackie JM, Dow JAT (Eds). (1989) The dictionary of cell biology. Academic Press, London, 262 pp.

Pierik RLM (1987) Abbreviations and glossary. pp. 7–20. In: In vitro culture of higher plants. Martinus Nijhoff, Dordrecht, 344 pp.

Rieger R, Michaelis A, Green MM (1991) Glossary of genetics—classical and molecular. 5th Edition. Springer Verlag, Berlin, Heidelberg, New York, 553 pp.

Snyder EB (1959) Glossary of forest tree improvement workers. Soc. Am. Foresters by the S. Forest Expt. Sta., Forest Service, USDA. 22 pp.

Subject Index